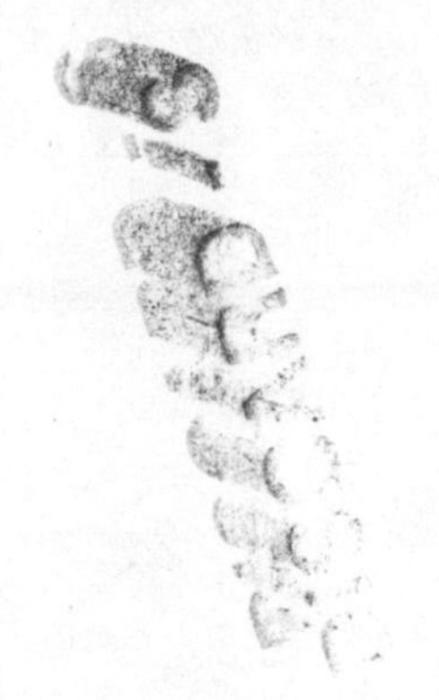

CELESTIAL NAVIGATION FOR YACHTSMEN

CELESTIAL NAVIGATION FOR YACHTSMEN

By
MARY BLEWITT

REVISED AMERICAN EDITION

Edited by

Frank G. Valenti, *Chief Quartermaster*

U.S. Coast Guard Reserve

JOHN de GRAFF, INC.
TUCKAHOE, NEW YORK

Copyright © by Mary Blewitt, 1964
Library of Congress Catalog Card Number 67-25097
ISBN 8286-0028-7

First published	1950
Fourth edition	1964
First American edition	1967
reprinted	1969
reprinted	1971

No part of this book may be reproduced in any form, by print, photoprint, microfilm or any other means without written permission from the publisher.

John de Graff Inc.
34 Oak Ave., Tuckahoe, N.Y. 10707

PRINTED IN U.S.A. BY SCIENCE PRESS, INC.

PREFACE

I fully realize my temerity in adding another book on navigation to the many excellent ones available. I have, however, met a large number of yachtsmen who would like to be able to take and work out a sight but have been discouraged from trying either by the complexity of the books, by the feeling that their mathematics are not up to standard, or by the annoying tendency of navigators in yachts to give the impression that celestial navigation is very difficult—a kind of black magic.

This is a book written for beginners by a beginner, and to encourage my readers I must admit here and now my mathematical shortcomings. I do not understand trigonometry, and for that reason there will be no mention of sines and cosines. My knowledge of geometry is very limited, but the only two propositions I find necessary are:

1. That two parallel lines crossing a straight line make equal angles; and
2. That if you know two sides and the included angle of a triangle the other side and the other two angles can be found.

Apart from this I can add and subtract and that is about all!

I have made no attempt to go into unnecessary details so that there are a number of slightly inaccurate statements. For instance, I say that one nautical mile equals one minute of arc on the Earth's surface: this is not strictly true, for the measurement varies at the poles and the equator, but this difference is never apparent in practical navigation. Also, for simplicity, I have left out any mention of such technicalities as the celestial equator or the celestial horizon.

The *Air Almanac* and H.O. 249 (*Sight Reduction Tables*

for Air Navigation), provide as complete and easy a method of working out a sight as anyone could wish. The haversine method is more accurate, but the errors from these tables are not large enough to affect yacht navigation, and they have one great advantage: by their simplicity they minimize the possibility of human error. When the navigator is tired and wet and the boat bouncing about, as is so often the case when a sight is seriously and urgently needed, there is less chance of making a mistake with these tables than with any other method. I have, therefore, taken it that my readers will be using these books and in the 'practical' part I have tried to explain the working of the various sights by this method and by no other.

In spite of my un-mathematical approach I have succeeded in taking sights with adequate accuracy even in a rough sea, and I can assure any beginner that the sense of triumph when a sight proves correct is well worth the effort involved. It is only the first step which is difficult. Take one sight and you will feel slightly bewildered, take two and the fog begins to clear, take a dozen and you will wonder what all the fuss was about.

I have aimed at simplicity throughout this book and can only hope that my explanations are clear enough to encourage a would-be navigator to 'go and take a sight' with an adequate idea of what he is trying to do and a modicum of confidence that he will be able to do it.

MARY BLEWITT

PREFACE TO THE REVISED AMERICAN EDITION

When I was asked to edit *Celestial Navigation For Yachtsmen,* for an American edition, I thought it would be simply a matter of a few editorial changes, since the British have a nice way with words and phrases.

I was surprised to find that many of the navigational terms used abroad differ in meaning considerably from those used here in the U.S.A. These changes have been incorporated in this new edition.

Many of the diagrams were re-drawn to conform to accepted practices here in this country, and many of the examples re-done and up-dated to suit both the cruising waters of the United States, and the recently revised format of the Air Almanac. Excerpts from all the tables necessary for working the sights in this edition have been added and grouped as appendices in the back of the book for easier reference. An appendix listing all the navigational terms and symbols used has also been included. It was felt that although the intent of the original British edition was to eliminate all unnecessary technicalities, (and still is, I hasten to say), the standard accepted navigational abbreviations and symbols, which, if one can learn to use and understand them, saves much time in working out sights. For this, and no other reason, were they included.

We now have a text well suited for any American who wishes to learn the art, if not the science, of celestial navigation.

Frank G. Valenti

Stamford, Connecticut

CONTENTS

Page

Preface 3
Preface to the Revised American Edition 5
Key to the Diagrams 9

SECTION ONE: THEORY 11
The Heavenly Bodies —— Geographical Position Declination —— Hour Angle —— Zenith —— Horizon —— Altitude —— Zenith Distance —— Azimuth Angle —— Great Circles —— Greenwich Mean Time Zone Time —— Zone Description —— Standard Time —— Daylight Saving Time —— Position Line ——Noon Sight —— Pole Star Sight —— Spherical Triangle —— Sextant and Observed Altitude —— Height of Eye —— Refraction —— Semi-Diameter —— Parallax

SECTION TWO: PRACTICE 39
Almanacs —— Tables —— Sun Sights —— Moon Sights —— Planet Sights —— Noon Sights —— Star Sights —— Pole Star Sights

SECTION THREE: NOTES 64
Sextant Errors —— Star Finding and Identification The Spherical Triangle —— Additional Tables South Latitudes

SECTION FOUR: APPENDICES 69
Appendix A: Navigation Abbreviations and Symbols ... 70
Appendix B: Excerpts from The Air Almanac 72
Appendix C: Excerpts from H.O. 249, Vol. I 83
Appendix D: Excerpts from H.O. 249, Vol. II 86
Appendix E: Excerpts from H.O. 249, Vol. III 88

INDEX 90

KEY TO THE DIAGRAMS

The key below applies to all the diagrams in this book.

P, P[1] .... North and South Poles.
E, E[1] .... Equator.
H, H[1] .... Horizon.
Q .... Center of the Earth.
X .... Geographical position of the heavenly body under discussion.
Z .... The observer.
Z′ .... Observer's zenith.
G .... Any point on the Greenwich meridian.

From the foregoing it follows that—

The line PZ is part of the observer's meridian.

The line PX is part of the meridian of the geographical position of the heavenly body under discussion.

The line PG is part of the Greenwich meridian.

Except where further description is necessary these letters are not explained again in the book.

SECTION ONE: THEORY

Before the theory of a sight can be understood there are certain facts about the Earth which must be thoroughly grasped, and certain terms which must be learned.

The heavenly bodies

We navigate by means of the Sun, the Moon, the planets and the stars. Forget the Earth spinning round the Sun with the motionless stars infinite distances away, and imagine that the Earth is the center of the universe and that all the heavenly bodies circle slowly round us, the stars keeping their relative positions while the Sun, Moon and planets change their positions in relation to each other and to the stars. This pre-Copernican outlook comes easily as we watch the heavenly bodies rise and set, and is a help in practical navigation.

Geographical position (GP)

At any moment of the day or night there is some spot on the Earth's surface which is directly underneath the Sun. This is the Sun's GP, and it lies where a line drawn from the center of the Earth to the Sun cuts the Earth's surface. It is shown in Fig. 1 at X. Not only the Sun but all heavenly bodies have GPs, and these positions can be found from the Almanac at any given moment. The GP is measured by declination and hour angle.

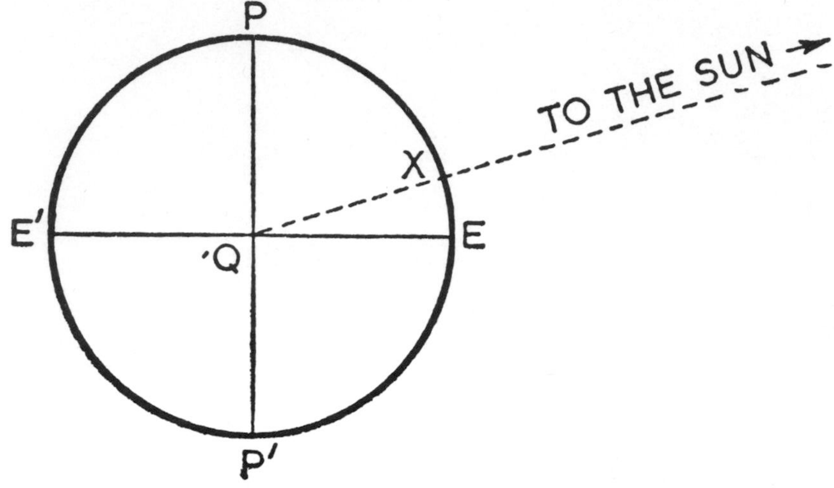

Fig. 1

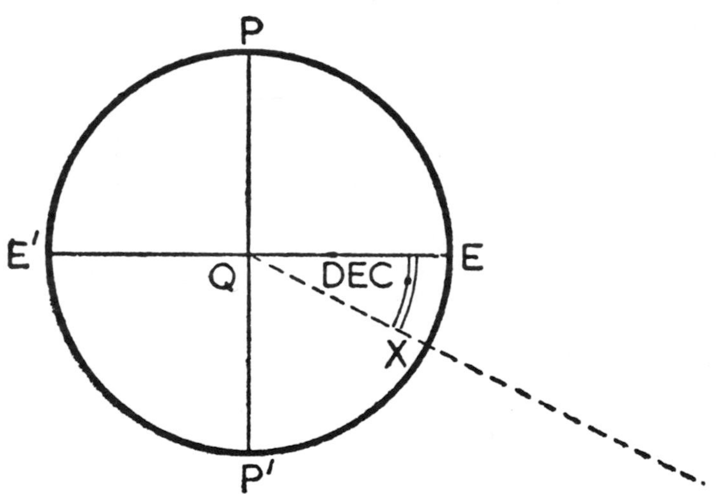

Fig. 2. Here ∠XQE is the declination of X; it is about 25° South

Declination (Fig. 2)

The declination of a heavenly body is the latitude of its GP, and is measured exactly as latitude, in degrees north or south of the equator. The declination of the Sun varies from 23° N. in midsummer when it reaches the Tropic of Cancer, to 23° S. in midwinter at the Tropic of Capricorn. In spring and autumn, at the equinoxes, it is 0° as it crosses the equator. The declination alters at an average rate of one degree every four days and is tabulated for every ten minutes in the *Air Almanac*.

The Moon's declination varies between approximately 28° 30′ N and S. It alters very quickly, sometimes as much as 6 or 7 degrees in 24 hours, and it also is tabulated for every ten minutes.

For the purposes of navigation the declination of stars can be considered as constant. The declinations of planets vary considerably and are given in the Almanac for every hour. No planet, however, has a declination greater than 29° N or S.

Hour angle

The GP of any heavenly body is not only on a parallel of latitude but also on a meridian of longitude, and hour angle is the method of measuring this meridian. It differs from longitude in some marked respects.

Let us consider the Sun. If you are standing on the Greenwich meridian in England at noon the Sun is due south of you and its hour angle is nil. Two hours after noon its hour angle is 2 hours. As the Sun sets, goes round the other side of the Earth and rises again, the hour angle increases until at eleven in the morning the hour angle is 23 hours while at noon it comes the full circle of 24 hours to start again at 0 as it crosses the meridian. The hour angle, when it is measured from the

Greenwich meridian, is called the Greenwich Hour Angle (GHA). It is always measured in a westerly direction and can be measured in time or in 'arc,' i.e., degrees, minutes and seconds (once round the Earth is 24 hours or 360°).

Unfortunately, you cannot measure this angle by your chronometer or watch because the Sun does not keep regular, or mean, time, and is sometimes as much as 20 minutes slow or fast by GMT, so that the angle has to be looked up in the Almanac where it is tabulated for every second of every day.

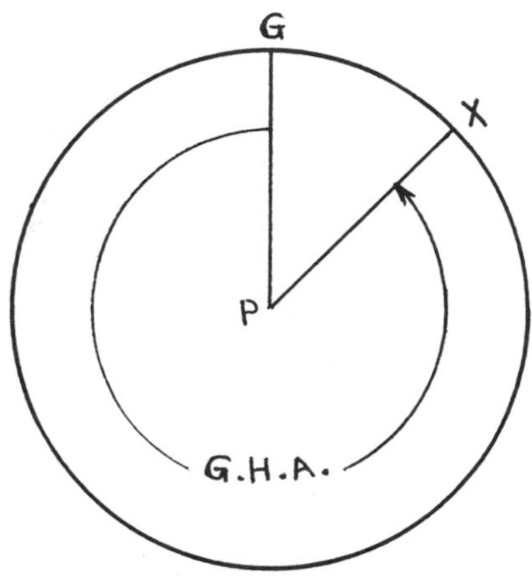

Fig. 3

In Fig. 3 we are out in space looking up at the South Pole. The GHA is measured west from the Greenwich meridian, as shown by the arrow. It is morning, for the Sun is coming up to the Greenwich meridian and the GHA is approximately 21 hours, or 315° (360° ÷ 24 = 15°, so 1 hour = 15°).

Now hour angle can be measured not only from the Greenwich meridian but from any meridian. When it is measured from the meridian on which you are standing it is known as

Local Hour Angle (LHA). This is also measured in a westerly direction. If you are west of Greenwich, LHA is less than GHA because the Sun passed Greenwich before it passed you, so the GHA is the larger angle. If you are east of Greenwich, LHA is greater than GHA since the Sun passed you first. Whereas GHA is found from the Almanac, LHA is found by adding or subtracting your longitude to or from the GHA.

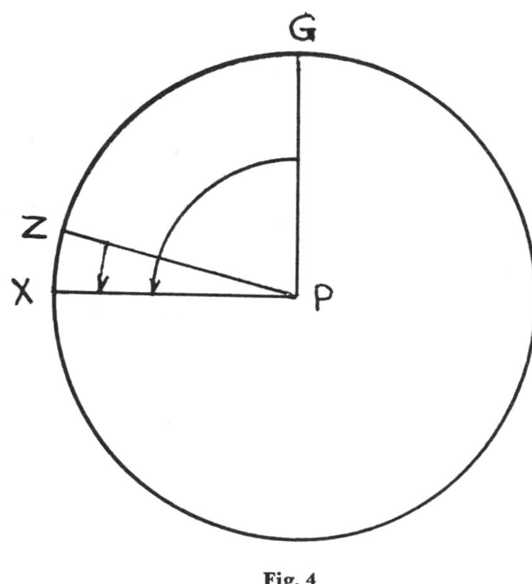

Fig. 4

Consider the following four examples:

Example A (Fig. 4). Suppose that you are north of the Bahamas (Long. 75° W) at one p.m. local time. As it is one hour after noon the Sun will be an hour past your meridian and the LHA (broken line) will be 1 hour (15°). But it is a long time since the Sun crossed the Greenwich meridian, so the GHA (unbroken line) will be much larger. It will be the 75°

of your longitude plus the 15° the Sun has gone past you, i.e., 90° (6 hours). In longitudes west: GHA – the observer's longitude = LHA. 90° – 75° = 15°.

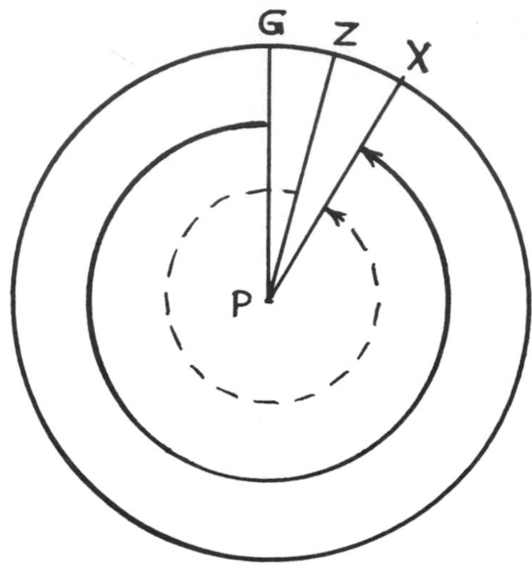

Fig. 5

EXAMPLE B (Fig. 5) You are somewhere in Italy (Long. 15° E.) at 11 a.m. local time. The LHA (broken line) is 23 hours or 345°, since it is an hour before your noon. The Sun has farther to go, however, to reach the Greenwich meridian so the GHA (unbroken line) is only 22 hours or 330°.

In longitudes east: GHA + observer's longitude = LHA. 330° + 15° = 345°.

EXAMPLE C (Fig. 6). You are somewhere in the Atlantic and the Sun has just passed the Greenwich meridian so that the GHA is small, say, 1 hour 30 minutes or 22° 30′; but the Sun has not yet reached your meridian PZ (Long. 52° 30′ W). On that meridian it is only 10 a.m. and the LHA will be 22 hours or 330°.

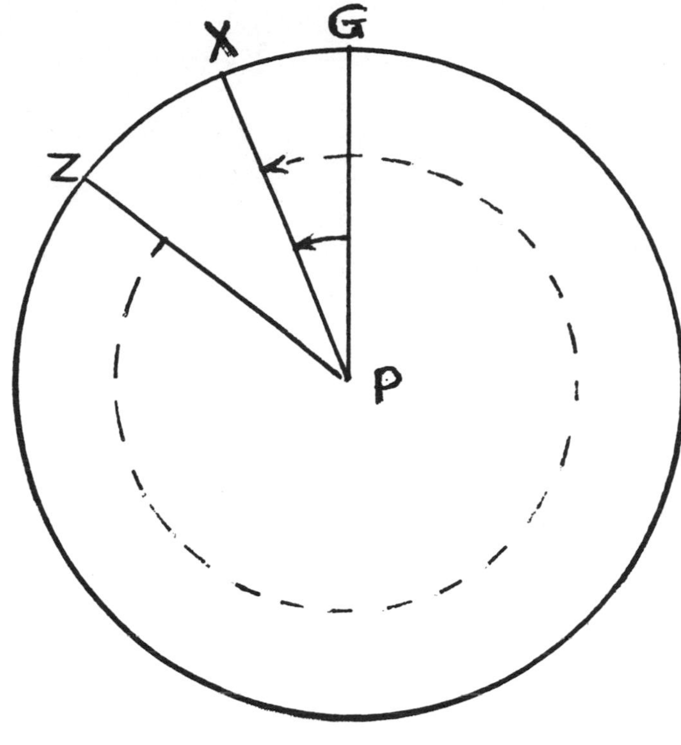

Fig. 6

The observer is west of Greenwich so again: GHA – observer's longitude = LHA. 22° 30′ – 52° 30′ = LHA. In this case you must add 360° in order to be able to do the subtraction.

$$
\begin{array}{r}
360^\circ\ 00' \\
+\ 22^\circ\ 30' \\
\hline
382^\circ\ 30' \\
-\ 52^\circ\ 30' \\
\hline
330^\circ\ 00'
\end{array} = \text{LHA.}
$$

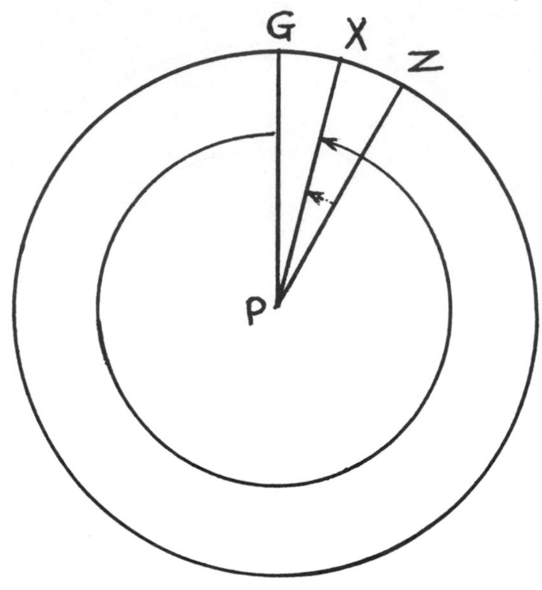

Fig. 7

Example D (Fig. 7) . Here you are in the eastern Mediterranean (Long. 30° E) . The Sun has passed your meridian but has not yet reached Greenwich. LHA therefore is small but the GHA is large, say, 345° or 23 hours. Now your longitude is east of Greenwich so GHA + observer's longitude = LHA. 345° + 30° = 375°, and 375° − 360° = 15° = LHA.

These diagrams apply not only to the Sun but to all heavenly bodies, but of course only the Sun crosses the Greenwich meridian at noon. For other bodies the time when the Almanac gives their GHA as 0° is the time of their meridian passage.

A thorough understanding of hour angle is essential, and it will be well to summarize what has been written so far.

First, hour angle differs from longitude in three main ways—

1. It may always be measured either in time or in arc (conversion either way is easily looked up in tables).
2. It is *always* measured in a westerly direction.
3. It may be:

   (*a*) GHA—measured from the Greenwich meridian.
   (*b*) LHA—measured from the meridian of the observer.
   (*c*) Sidereal Hour Angle (SHA). This is explained later on page 56.

Secondly, the GHA of any heavenly body can be found from the Almanac for any given moment. LHA is obtained in longitudes west by subtracting the observer's longitude from GHA, in longitudes east by adding it to the GHA.

We have now seen that the GP of a heavenly body is determined by declination and hour angle, and that at any given moment this GP could be plotted on a chart, although in fact it is never necessary to do so.

Zenith

If a line were drawn from the center of the Earth through you and out into space it would lead to your zenith. In other words, it is the point in space directly above your head. For instance, if you were standing on the GP of the Sun then the Sun would be in your zenith.

Horizon

As it is impossible to see round a corner we cannot see much of the surface of the Earth, which bends away from us in all directions. The horizon lies in a plane which at sea level is at a tangent to the Earth's surface, and that plane forms a right angle with the direction of the observer's zenith.

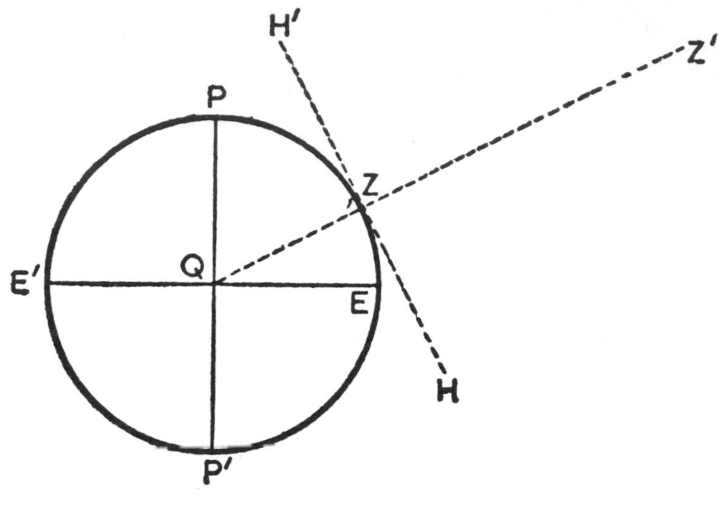

Fig. 8

In Fig. 8 HH′ is a tangent to the Earth's surface at Z and both ∠ Z′ZH and ∠ Z′ZH′ are right angles.

Altitude

The altitude is the angle made at the observer between the Sun (or any other heavenly body) and the horizon directly below it. In Fig. 9 ∠ HZS is the altitude of the Sun. This is the angle you measure with a sextant when you take a sight.

Zenith distance

The zenith distance is the complement of the altitude. In Fig. 9, it is ∠ Z′ZS. Altitude + zenith distance *always* equals 90°.

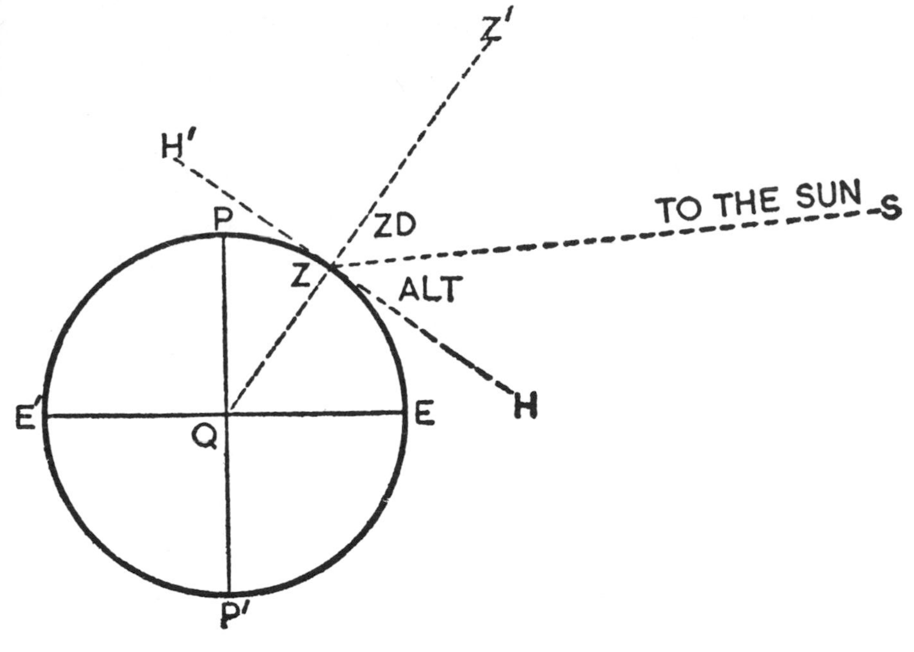

Fig. 9

Azimuth angle

This is the bearing (true, *not* magnetic) of any heavenly body (and of its GP) from the observer. It is measured from the north (in the northern hemisphere) through east or west: from N through 180° E and from N through 180° W. Fig. 10 shows various azimuth angle readings. When working out a sight, azimuth angle is obtained from the tables where a single figure is found. Whether this is N 145° E or N 145° W depends on whether the heavenly body has passed your meridian or not. In the morning the azimuth angle of the Sun will be N and E, in the afternoon N and W. To convert azimuth angle (Z) to true azimuth (Zn), i.e. to the 360° system found on the compass, the rule in north latitudes is as follows:

LHA greater than 180°: Azimuth angle (Z) = True azimuth (Zn)

LHA less than 180°: 360 — Azimuth angle (Z) = True azimuth (Zn)

For instance, LHA 22°, azimuth angle 145° ∴ 360 — 145 = 215°. Had you looked along your compass when you took your sight (allowing for magnetic variation, of course) the heavenly body would have been on a bearing of 215° from you.

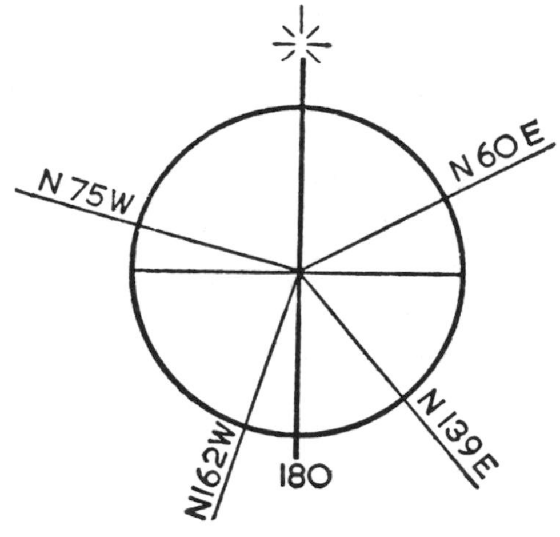

Fig. 10

Great circles

A great circle is any circle with its center the center of the Earth and its radius the distance from the center to the surface of the Earth. The equator and the meridians are great circles, but parallels of latitude, except the equator, are not, because the center of a circle formed by a parallel of latitude lies either north or south of the center of the Earth. Distances along great circles can be measured in two ways, by miles or by the angle subtended at the Earth's center.

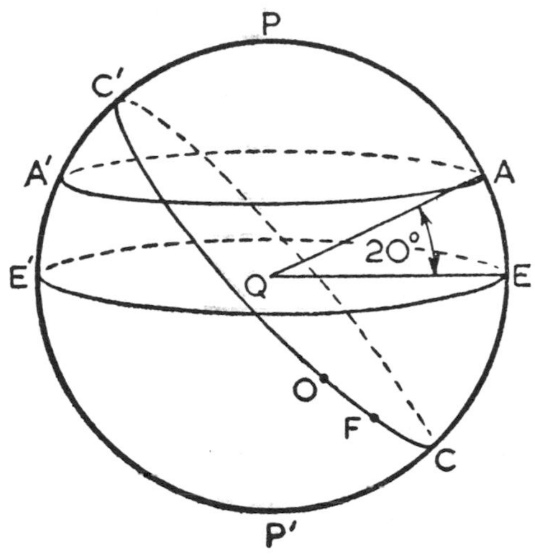

Fig. 11

In Fig. 11 three great circles are shown. EE′, CC′ and the circle PEP′E′. AA′ is a parallel of latitude but not a great circle. The distance AE can be measured in arc. It makes an angle of 20° at Q. Or it can be measured in miles, and, since one minute of arc on the Earth's surface equals one nautical mile, AE = 20 × 60 = 1,200 miles. CC′ is a great circle (although it is not a meridian, for it does not go through the poles). Therefore the distance between O and F, two points on that circle, could also be measured in arc or in miles.

This method of measurement can be used between any two points on the Earth's surface (the shortest distance between any two points being part of a great circle) . This interchangeability of arc and mileage should always be kept in mind.

Time

There is no necessity for the beginner to plunge into the intricacies of time, which are many, but he must understand Greenwich Mean Time, Zone Time, Zone Description, Standard Time, and Daylight Saving Time.

Greenwich Mean Time

GMT is the time given to the world by the Greenwich Observatory. It is an average, or mean, of the Sun's time because noon GMT is rarely the time at which the Sun actually crosses the Greenwich meridian. The Sun may be as much as 16.4 minutes before or after noon.

Zone Time

ZT is the time kept, for instance by the passengers and crew of a big ship at sea and changes one hour for every change of 15 degrees of longitude.

Zone Description

ZD is the number, with its sign (+ or —) that must be added to, or subtracted from, zone time to obtain Greenwich mean time. The zone description is usually a whole number of hours. In west longitude the ZD is (+) ; in east longitude it is (—) .

Standard Time

This is the time laid down by various countries for general use. It often coincides with zone time, but not always. India, for example, which lies in zones 5 and 6 hours east of Greenwich, uses a standard time of 5 hours 30 minutes for the whole country. A list of standard times is given in the *Air Almanac.*

Daylight Saving Time

DST is the time used during the summer so as to obtain an hour's more light in the evening. One hour is added to standard time to obtain DST.

Before going on I advise readers to whom all this is new to re-read what has been written so far, because it is important to understand it thoroughly before continuing.

The Position Line (Fig. 12)

The final result obtained from any sight on any heavenly

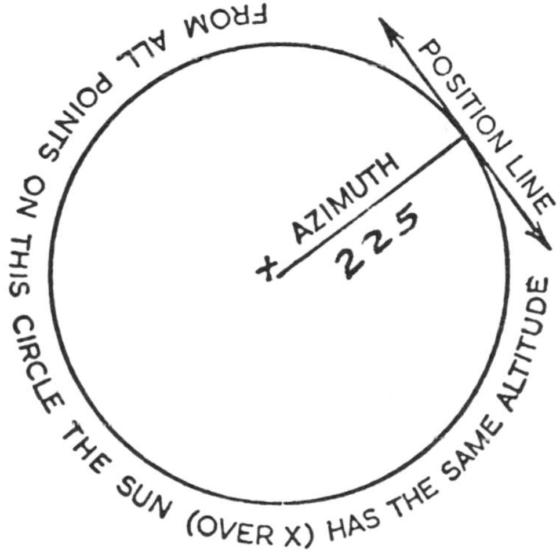

Fig. 12

body is a straight line on your chart and you are somewhere on that line. If the Sun is in your zenith its altitude is 90°, and there is only one spot on the Earth's surface where you can be—at the GP of the Sun. As you move away from the GP the altitude will lessen, and it will lessen equally whether you go north, south, east or west. Move away until the altitude of the Sun is 85°, and you are on part of a circle with its center at the GP. At every point on that circle the altitude will be 85°.

If the Sun is SW of you you will, of course, be on the NE portion of the circle, but unfortunately it is impossible to obtain the azimuth of the Sun accurately enough to fix your exact position on the circle. The only thing to do is to draw a line at right angles to the most accurate azimuth you can get and say, 'I am somewhere on this line'. The line is drawn straight because the distance from the center of the circle is so great that it is impracticable to show the curve of the circle on the chart.

It is helpful to realize how very large these 'position' circles are. For example, in the winter in the morning when the Sun is over SW Africa, and its altitude in England is about 12°, the Sun also has an altitude of 12° near the following places: Greenwich, the Caspian Sea, Madras, the South Pole, Chile, British Guiana and the Azores. Even in midsummer at noon in New York, when the Sun is north of Cuba, and at its highest (72°), the circle runs through New York, along the Mississippi, across the Gulf of Mexico, south of Panama, past Bogota, through Trinidad, and over the Atlantic back to New York.

We have seen that the position line is at right angles to the azimuth and this is of practical value, not only for working out sights but also for determining when is the best time to take them. In Fig. 13, you are approaching a strange coast from the NW and are not certain of your position. A morning sight of the Sun when it is SE might give you the line AA′ which will determine your distance off land, while from a sight in the afternoon you have the line BB′ which positions you along the coast.

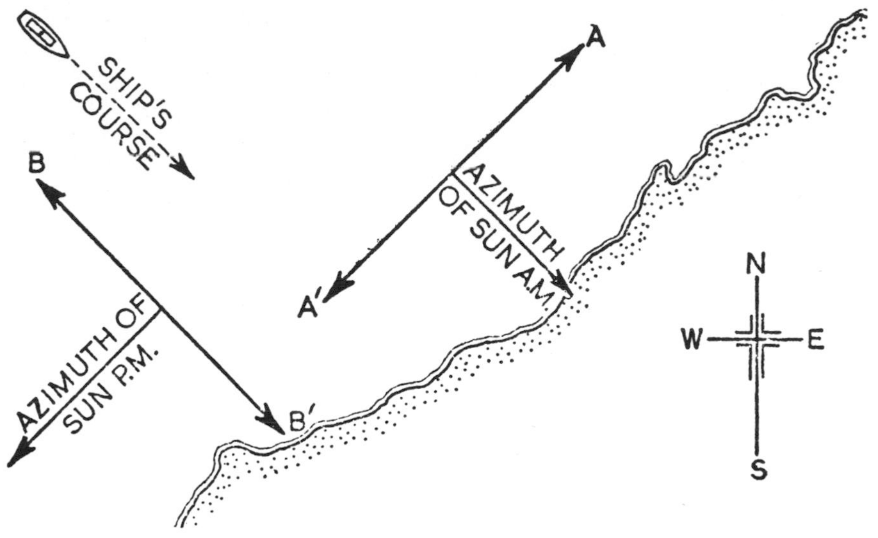

Fig. 13

Noon Sight

A position line is at right angles to the bearing of any heavly body, so when a 'body' is on your meridian, that is, when it is due north or south of you, your position line will run east and west, and a line running east and west is a parallel of latitude. This has a special application to the Sun as it is the principle on which a 'noon sight' is worked.

In Fig. 14, PZXP′ is the meridian on which both Z and X lie. All the heavenly bodies are so far from the Earth that their rays strike the Earth in parallel lines. (This, however, is subject to corrections discussed on page 37.) We have here, then, two parallel lines, SZ and S′XQ, crossing the straight line Z′ZQ. Therefore, $\angle Z'ZS = \angle ZQX$. Now Z′ZS is the zenith distance, so that if we take a sight when the Sun is due south we know the size of these angles $(ZD = 90° - \text{Alt.})$. From

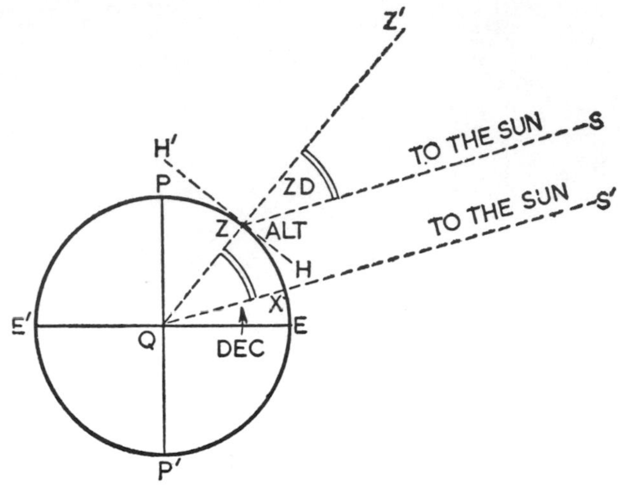
Z'
H'
P
ZD
TO THE SUN
S
Z
ALT
TO THE SUN
S'
H
X
E'
E
Q
DEC
P'

Fig. 14

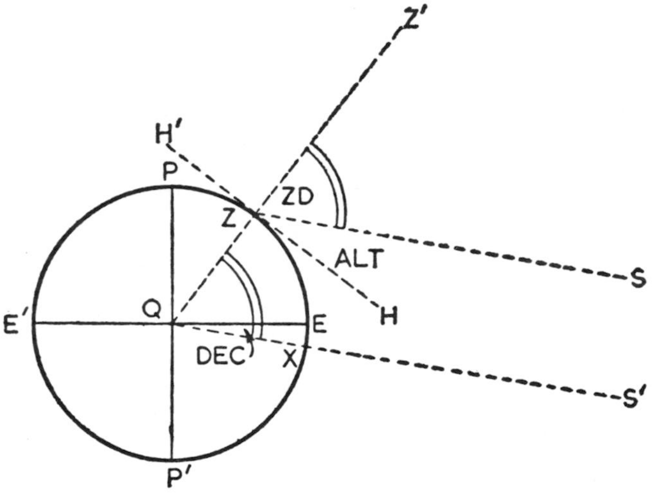
Z'
H'
P
ZD
Z
ALT
S
Q
H
E'
E
DEC
X
S'
P'

Fig. 15

the Almanac we know ∠XQE, for it is the declination. Add ∠ZQX and ∠XQE and the resulting angle ∠ZQE is the latitude of Z. This is to say: zenith distance + declination = latitude.

This applies with a north declination. Fig. 15 shows the noon sight with a southerly declination, and in this diagram it will be seen that ∠ZQX includes ∠EQX and that the latitude of Z = ∠ZQX − ∠EQX. That is, zenith distance − declination = latitude.

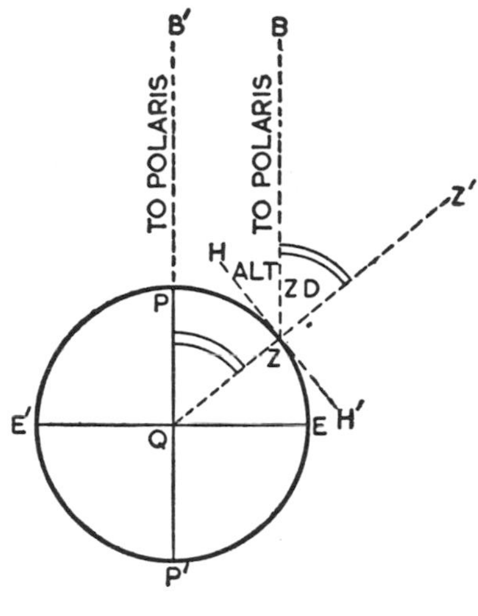

Fig. 16

Pole Star Sight (Fig. 16)

The same principle applies to the Pole Star sight although certain corrections have to be used as *Polaris* is not exactly over the North Pole. Presume, however, for the moment that it is, that its GP is always the North Pole. Then it is always due north of you and its position line will always be a parallel of latitude.

In Fig. 16, P marks the GP of *Polaris* as well as the North Pole because they are the same point, and B and B′ mark the direction of *Polaris*. Now ∠BZZ′ = ∠PQZ, but they are both parts of the right angles, ∠HZZ′ and ∠PQE, so that ∠HZB = ∠ZQE. ∠HZB is the altitude of *Polaris* and ∠ZQE is the observer's latitude. Therefore the altitude of *Polaris* equals the latitude of the observer.

Spherical Triangle

We have seen that in each of the last three diagrams the zenith distance is equal to the angle subtended at the Earth's center by the observer and the GP. This is true not only of noon sights or *Polaris* sights, when the GP is on your meridian, but of all sights, whatever the bearing of the heavenly body. The zenith distance *always* equals the angle at the Earth's center made by the GP and the observer. (In Figs. 14 and 15, ∠ZQX will always equal the zenith distance even if the line ZX is not part of a meridian but part of some other great circle.) This angle can, of course, be translated into miles at the Earth's surface. An altitude of 47° gives a zenith distance of 43° which, since 1′ of arc = 1 nautical mile, means that the GP is (43 × 60) 2,580 miles away from the observer.

We now know that by taking a sight and finding the zenith distance we can find our distance from the GP of any visible heavenly body, but, because the distances involved are so great, we cannot put a compass on the GP and draw the required circle. Nor can we mark the position line on our chart except when the body is on our meridian and the distances are conveniently marked by parallels of latitude.

We must therefore approach the problem from a different direction. We pretend that we *do* know where we are but that we do *not* know the altitude of the 'body'. In Fig. 17 we are pretending that we know our latitude and longitude, that is we work from an *assumed* position.

In Fig. 17 we are looking at the outside of the Earth. PAP′ and PBP′ are the meridians of X and Z respectively, crossing the equator at A and B. What do we know about the triangle PZX? We know the length of the side PZ. BZ is the observer's latitude, therefore PZ = 90° – latitude. We know the length of the side PX since AX is the declination: therefore PX = 90° – declination. And we know the included angle ∠ZPX since this is the local hour angle (the angle between the ob-

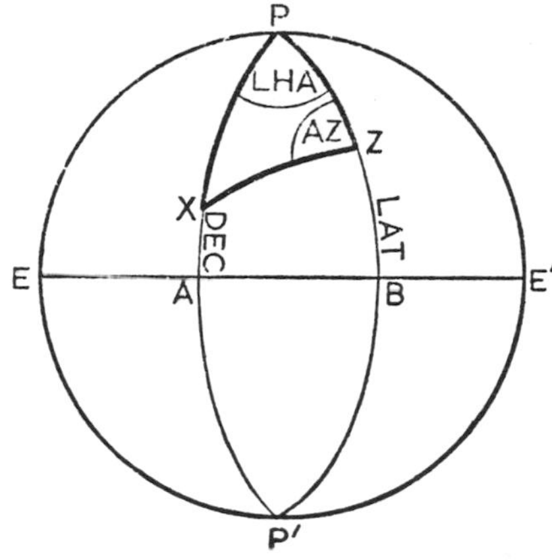

Fig. 17

server's meridian and the meridian of the GP measured in a westerly direction).

Now if we know two sides and the included angle the triangle can be solved, and by spherical trigonometry we can find out the length of the side XZ and the other two angles. The length of the side XZ is equal to the zenith distance, so that 90° – XZ = altitude. Since it is looked up in tables it is called the *computed* altitude. The computed altitude is the same as the altitude we should have gotten from our sextant if we had taken a sight at our assumed position at that particular time.

Suppose that on 18 August 1966, at 14:43:22 GMT, we took a sight of the Sun and got an observed altitude of 41° 38′. We knew that we were somewhere SW of the Isle of Wight and we assumed that our position was 50° 00′ N 01° 54′ W (the reasons why we chose this position will become apparent later.) After the necessary calculations we get a computed altitude from the tables of 41° 43′; that is to say, if we had been at the above position at that time our observed altitude would have been 41° 43′, too, but it was not, so we were somewhere else. The difference between the computed altitude and the observed altitude is 5′, and since 5′ = 5 miles our position line will be 5 miles away from our assumed position. This is called an *intercept* of 5 miles.

If you look again at Fig. 17 you will see that ∠PZX is the azimuth, and as so much is known about the triangle this angle can be worked out. In the sight we have just taken the tables tell us that the azimuth angle was 127° and because it was afternoon we know this was N. 127° W.

On our chart we draw the true azimuth (360 − 127 = 233) through our assumed position. As has already been explained, the position line lies at right angles to the azimuth, our intercept was 5 miles, so our position line will be 5 miles 'away' from or 'towards' the Sun from our assumed position. The farther away from the GP we get, the less the altitude, so if our observed altitude were less than our computed altitude we must have been farther away than we assumed. In this case the computed altitude was 41° 43′, the observed altitude 41° 38′, so the position line will be 'away' from the Sun. Fig. 18 shows the working on the chart.

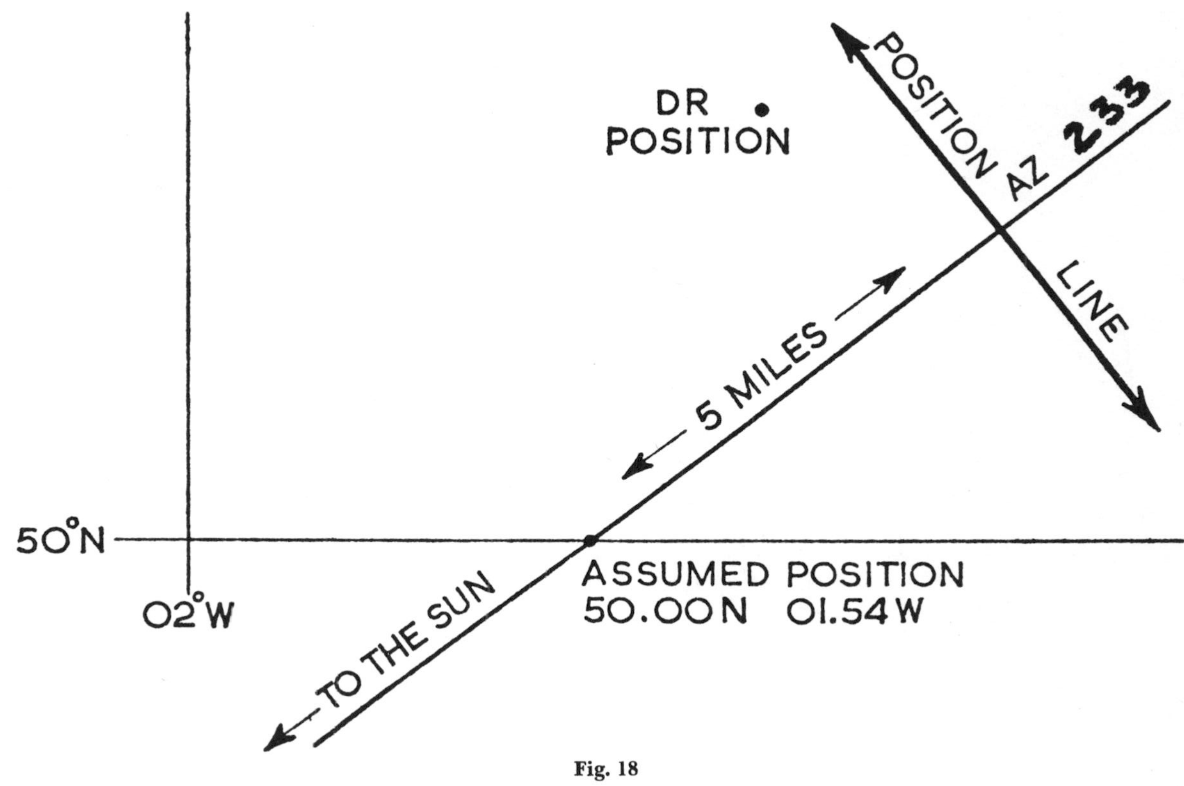

DR POSITION
POSITION AZ 233 LINE
5 MILES
50°N
02°W
ASSUMED POSITION
50.00N 01.54W
TO THE SUN

Fig. 18

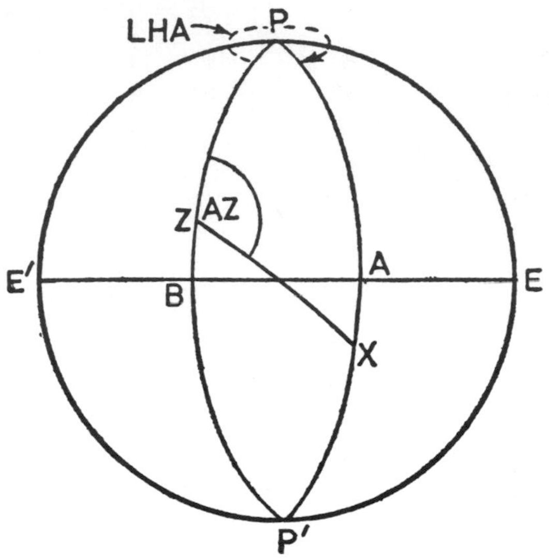

Fig. 19

Fig. 19 shows another example of the triangle which always has to be solved. In this diagram X is the GP of the Sun. From merely looking at this triangle we can tell certain things: it is morning at Z, for the Sun is to the east, and it is winter, for the Sun's declination is south of the equator.

Now, PZ = 90° – latitude of observer. PX = 90° + declination and ∠ZPX = LHA. The fact that here our LHA will be, say, 323°—being measured west from the meridian of Z—and that ∠ZPX is obviously (360° – 323°) 37° need not worry us, as it is all done by the tables. The important point is that we know the size of the angle.

ZX can now be calculated and the computed altitude (90° – ZX) found in the tables. The azimuth ∠PZX, can also be calculated and found in the tables.

This is the principle of every sight taken—that with the

known facts about a heavenly body and an assumed position, we can calculate an altitude, and that as the computed altitude varies from our observed altitude so our position line varies from our assumed position.

Sextant and Observed altitude

When you take a sight the angle you read off from your sextant is called the *sextant* altitude, and there are certain corrections which have to be applied to the sextant altitude in order to obtain an *observed* altitude for comparison with the computed altitude. These briefly are as follows:

1. Height of eye (dip of sea horizon)

The computed altitude from the tables is worked out as if you were at sea level, but taking a sight from a boat your height of eye may be 6 or 60 feet above sea level, depending on whether you are in a yacht or a liner, and a correction for this must be made to your sextant altitude to make it a sea level reading. (In small boats it is usually about — 3′) . This correction must be applied to all sights except those taken with a bubble sextant or an artificial horizon. The dip table appears on the outside of the back cover of the *Air Almanac,* (see page 82).

2. Refraction

This is a correction for the bending of the rays of light by the Earth's atmosphere. It is applied to *all* sights. The smaller the altitude of a heavenly body, the greater the refraction. Sights are inadvisable below 6° and should be treated with caution below 10° of altitude. The refraction table is on the inside of the back cover of the *Air Almanac,* (see page 81). It is drawn up for aircraft and only the first vertical column applies to ships. (If you are using the *Astronomical Navigation Tables* (H.O. 218) see page 68).

3. Semi-diameter (Fig. 20)

This correction applies only to the Sun and Moon. In theory, when taking a sight the horizon should bisect the 'body'. This, however, is impracticable. When taking the Sun the bottom or lower limb is usually rested on the horizon. The upper limb of the Sun can, of course, also be used if the lower limb is obscured by cloud. With the Moon the upper or lower limb may be used, depending on which is available. In each case half the diameter of the body must be allowed for. Semi-diameter is given in the Almanac each day for both the Sun and the Moon. The correction, usually 16′, is *added* for the

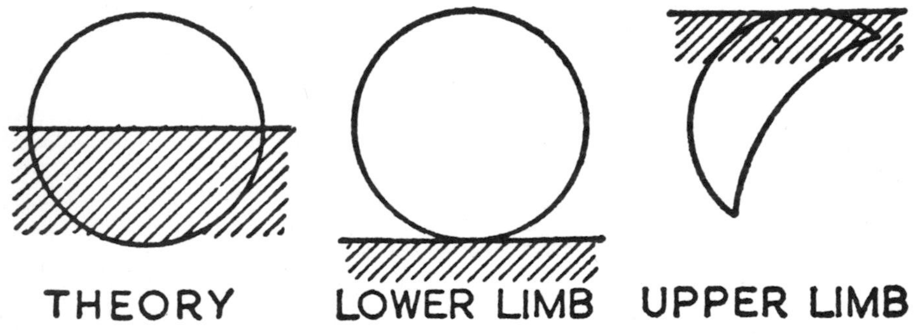

Fig. 20

lower limb and *subtracted* for the upper limb. It is not applied if you are using a bubble sextant or an artificial horizon.

4. Parallax

It has been mentioned earlier that the rays from heavenly bodies strike the Earth in parallel lines. This is not strictly true, but as far as the Sun, stars and planets are concerned the correction for parallax is negligible and can be ignored. The Moon, however, is so much closer to the Earth that a considerable correction is usually necessary, and this is tabulated each day in the Almanac.

In addition to these corrections there may also be sextant corrections for errors in the instrument itself, and these are dealt with later under Notes on Sextants. The tables in the back of the *Air Almanac* for dome refraction and coriolis apply to aircraft only.

SECTION TWO: PRACTICE

Two kinds of books are needed for working out sights:

1. Almanacs

These are temporary books which must be renewed every year. They give the positions of the heavenly bodies for every minute of every day throughout the year, as well as additional information such as sunrise and sunset, the Moon's phases, eclipses, standard time throughout the world, etc.

You need only one almanac, but there are two from which to choose:

The *Air Almanac* is used for the examples in this book and is suitable for use with H.O. 249 (see below). It appears in three volumes: January–April; May–August; and September –December.

The *Nautical Almanac* is based on the same principles as the *Air Almanac*, but it is slightly more detailed and more complicated. If, however, you are accustomed to the *Air Almanac* you will find no difficulty in changing over to the *Nautical Almanac*. It is published in one volume for the year and is suitable for use with H.O. 214 (see below).

2. Tables

These are books which do not go out of date, with the one

exception of H.O. 249, Vol. I, which is republished every ten years. Tables provide tabular solutions of the nautical spherical triangle, i.e. they enable you to find the included angle and the third side of the spherical triangle without doing any work, once the necessary figures have been obtained from the Almanac.

There is a large choice of tables, but the easiest to use are:

1. Sight Reduction Tables for Air Navigation (H.O. 249)

These tables, from which the examples in this book are taken, consist of three large volumes:

Vol. I Selected Stars (Epoch 1965.0)
Vol. II Latitudes 0° — 39° Declinations 0°—29°
Vol. III Latitudes 40° — 89° Declinations 0° — 29°

2. Tables of Computed Altitude and Azimuth (H.O. 214)

Vols. I through IX. See page 68.

3. Astronomical Navigation Tables (H.O. 218)

See page 68.

SUN SIGHTS

Let us pretend you are off the coast of Delaware, DR position 38°25′N, 74°21′W, on the afternoon of 1 October 1966, and you wish to take a sight of the Sun.

In order to work out your sight you need two things: the sextant altitude of the Sun and the exact time at which you took that altitude. Do not try at first to take your own time. Get someone else to do it for you—and make sure that he can tell the time! It is quite amazing how many people there are who, when you say "mark," do not write down the correct time. A mistake of only four seconds may lead to an error of one nautical mile.

Take your sextant and make yourself comfortable, well supported and firm from the waist down and mobile from the waist up. Make sure you have a clear view of the Sun and of the horizon below it. Generally, the higher you are the better, because you can more easily avoid false horizons due to wave tops, but it is not worth while making yourself unsteady to gain height. It is also obviously important to choose a place where there is a minimum of spray. Arrange your shade glasses so that you have a good clear Sun which will not dazzle you and use a very light horizon glass if necessary (usually it is not).

The clarity of the horizon is of the greatest importance. Sometimes in calms there is not enough difference between air and water to distinguish the horizon clearly. On other occasions mist obscures the real horizon and gives a false horizon nearer to you. Very rarely there may be abnormal refraction. This can sometimes be noticed because the horizon, seen through the sextant, may appear to 'boil', or the funnels of distant ships appear to reach up into the sky. If you have any doubts about the horizon, if it looks in any way odd, or hazy, either don't take your sights, or treat them with caution.

When you have the Sun more or less on the horizon, rock the sextant gently from side to side and you will see the Sun swing as if it were attached to a pendulum. It is at the lowest point of this swing that you take the sight.

Do not try to take a sight of any heavenly body with an altitude of 65°, or at most 70°. It is very difficult to judge where on the horizon the Sun, or other body, is at its lowest and you may get a considerable error.

When you feel sure that the Sun is just resting on the horizon, call to your time-keeper and then read off to him the degrees and minutes from your sextant to the nearest half-minute of arc. Take a series of five sights at about one-minute intervals, or quicker if and when you can. They will appear something like the following:

SUN Saturday, 1 October 1966

	Watch Time			Sextant
	h	m	s	° ′
1	14	11	55	36 31.′5
2	14	12	28	36 26
3	14	12	52	36 23
4	14	13	16	36 18.′5
5	14	13	40	36 12

Now average these sights and times. This, in my opinion, is the most tiresome job in the whole observation, but remember that the more sights you include in your average, the less likely the chance of a large error.

Averaging gives a watch time of 14:12:50, and a sextant altitude of 36°22′. With these figures you can start to work out your sight.

1. Watch Time to GMT

	h	m	s	
Watch Time (WT)	14	12	50	(Do not forget Daylight Saving Time. If in force subtract 1 hour)
Watch Error (WE) (fast)	−		5	
Zone Time (ZT)	14	12	45	
Zone Description (ZD)	+ 5	00	00	
GMT	19h	12m	45s	

2. Sextant Altitude to Observed Altitude (see page 35)

Sextant altitude (hs)	36°22′	
Index error (IE)	— 2	For IE see notes on sextants, page 65.
Dip (D) (8 ft.)	— 3	
Refraction (R)	— 1	
Semi-diameter (SD)	+ 16	SD is found in the bottom right-hand corner of every page of the Almanac.
Observed altitude (Ho)	36°32′	

3. LHA Sun from GMT

In the *Air Almanac* every day has a whole leaf which can be torn out and thrown away when the day is over. The first side is for midnight until noon (Greenwich A.M.) and the second from noon until midnight (Greenwich P.M.). On the page for 1 October 1966 (page 75) under ☉ Sun, you will find the GHA given for every 10 minutes.

For 1910 the GHA is 110°04′.6, and this should be rounded off to 05′ (otherwise you must use the interpolation table on page A80 of the Almanac). You are left with 2 minutes 45 seconds unaccounted for. Inside the front cover of the Almanac and on the outside of the folded flap at the end of the daily pages is a table headed 'Interpolation of GHA'. (See page 76).

This has three columns, the two outside ones are headed SUN Etc., and MOON, with times from 0 to 10 minutes beneath them, the center column of arc is common to both and gives the interpolation of GHA for the minutes and seconds of time over and above an exact 10 minutes. The figures from this column are *always* added. Looking at 2 minutes 45 seconds under the SUN column, we find against it 41′ (when you get an exact figure take the upper value). So you arrive at this:

	h	m	s	
GHA for	19	10	00	100° 05′
Increment for		2	45	41
GHA Sun for	19h	12m	45s	100° 46′

Now the included angle of the spherical triangle was LHA, not GHA, and when discussing hour angles we saw that LHA = GHA ± longitude. The easiest way to remember which is which is the old rhyme,

> Longitude east, GHA least,
> Longitude west, GHA best.

In this case your DR longitude is 74°21′W, so you must subtract your longitude to get LHA.

It becomes necessary at this point to explain the 'assumed position'. In the simplified methods of navigation you work not from your DR position but from an assumed position determined by the following three rules:

1. Your assumed position must be as near your DR position as possible.

2. Your assumed latitude must be a whole number of degrees.

3. Your assumed longitude must be so arranged as to make your LHA a whole number of degrees.

Rules 1 and 2 are quite simple and need no explanation. Rule 3 is a little more difficult and three imaginary cases are given below to show how it works:

(*a*) DR Long. 04° 50′ W

		°	′	
GHA		27	32	
Assumed Long (*a*λ)	−	04	32	(west subtract)
LHA		23	00	

(*b*) DR Long. 08° 25′ E

	°	′	
GHA	337	01	
Assumed Long (*a*λ)	+ 07	59	(east add)
LHA	345	00	

(c) DR Longitude 150° 36′ E.

GHA	314°	19′	
Assumed Long (*a*λ)	+150	41	(East add)
	465	00	(When this figure is more than 360, subtract 360 to arrive at LHA)
	−360	00	
LHA	105°	00′	

Let us go back to our sight. The DR longitude is 74°21′W, and GHA 110°46′. To get LHA, therefore, we have:

GHA	110°	46′
Assumed Long (*a*λ) (W) −	74	46
LHA	36°	00′

If the assumed longitude is a larger figure than GHA, 360 must be added to GHA to make the subtraction possible.

4. Declination

On page 75 you will see on the right-hand side of the Sun column the declination of the Sun given for every 10 minutes. At 1910 it is S 03°14.1. This figure should also be rounded off to the nearest minute of arc so we call it S 03°14′.

5. Assumed latitude

As has been said, your assumed latitude must be an integral degree, and since your DR latitude was 38°25′N, it is obvious that your assumed latitude will be 38°N.

You now have the three essentials for entering H.O. 249:

LHA	36°
Assumed Lat. (*a* L)	38°N
Declination (Dec.)	03°14′S

In H. O. 249, Vol. II (Latitudes 0 - 39, Declinations 0 - 29), turn to latitude 38. You will find four headings: 'Declination (0 - 14) *same* name as latitude,' 'Declination (0 - 14) *contrary* name to latitude' and two similar headings for declinations 15 - 29. *Name* in this context refers to 'north' or 'south'. In this example we must look under *contrary* since latitude is north and declination south. (See page 86).

Declination is given across the top and bottom of the tables and LHA vertically at the sides. Against LHA 36 and under declination 3° you find 37°11′ — 49′ 133° (see page 86). Now 37°11′ is the tabulated altitude (Hc at the top of the column stands for 'altitude computed') for a declination of 03°, but the Sun's declination when we took the sight was 03°14′, so the altitude must be corrected for 14′. At the back of H. O. 249, and on a loose card, you will find 'Table 4—Correction to tabulated altitude for minutes of declination'. (See page 87). Under 49 (the figure obtained from column *d*) and against 14 you find 11′. These 11′ must be subtracted as there is a minus sign before the 49 (correspondingly added when there is a plus sign) from 37°11′ to get your computed altitude:

$$\begin{array}{r} 37^\circ\ 11' \\ -\quad 11' \\ \hline 37^\circ\ 00' \end{array}$$

The last figure from the tables, under Z, is the azimuth angle, and since the Sun has already passed your meridian it is N 133 W, or a true azimuth (Zn) of 227°. Now your observed altitude was 36°32′ and your computed altitude is 37°00′ so there is a difference, or *intercept,* of 28 miles. This will be *away* from the Sun because the computed altitude is greater.

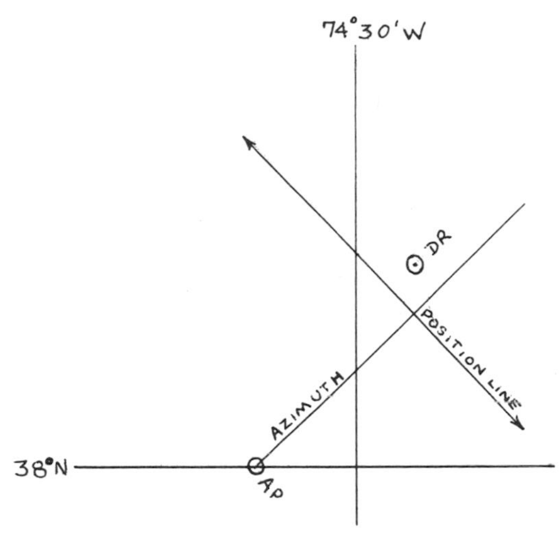

Fig. 21

This is clear when you think it out for if the angle were more you would be more underneath the Sun and therefore nearer to it, but as it is less you must be further away. In practice, however, I advise writing down 'Computed greater away' where you can always see it and then go by rule of thumb. You are now ready to put the position line on the chart. Mark the assumed position, draw your azimuth from it (true *not* magnetic), mark off 28 miles away from the Sun, draw a line at right-angles to the azimuth, and there you are! (Fig. 21).

In case this seems very complicated the figures used are given below without any of the explanation:

SUN Saturday, 1 October 1966. DR Lat. 38°25′N. Long. 74°21W.

	h	m	s
Watch Time (WT)	14	12	50
Watch Error (WE) (f)	–		05
Zone Time (ZT)	14	12	45
Zone description (ZD)	+ 5	00	00
Greenwich Mean Time	19h	12m	45s

Sextant altitude (hs)	36° 22′	
Index error (IE)	– 2	
Dip (D)	– 3	from the Almanac
Refraction (R)	– 1	from the Almanac
Semi-diameter (SD)	+ 16	from the Almanac
Observed Altitude (Ho)	36° 32′	

	h	m	s		
GHA for	19	10	00	110° 05′	from the Almanac
Increment for		2	45	41	from the Almanac
GHA Sun for	19h	12m	45s	110° 46′	
Assumed Long. ($a\lambda$) (W)				– 74 46	
LHA Sun				36° 00′	

Declination (Dec.) 03°14′S (from Almanac)

Assumed Latitude (*a*L) 38°N

Tables	Hc	*d*	Z
Tabulation	37° 11′	–49	133
d correction	– 11		
Computed Alt. (Hc)	37° 00′		
Observed Alt. (Ho)	36° 32′		
Intercept (*a*)	28 mi. AWAY (A)		

Azimuth Angle (Z) N 133° W
True Azimuth (Zn) 227°

This is the total work involved, from deciding on your average sight to putting it on the chart!

Let us take another Sun sight.

SUN Friday, 9 September 1966. DR Lat. 40°55′N. Long. 70°56′W.

	h	m	s
Watch Time (WT)	10	57	46
Watch Error (WE) (s)	+		52
Zone Time (ZT)	10	58	38
Zone description (ZD)	+ 5	00	00
	15	58	38
Daylight Saving Time (DST)	− 1	00	00
Greenwich Mean Time	14h	58m	38s

Sextant altitude (hs)		47° 33′
Index error (IE)	−	2
Dip (D)	−	3
Refraction (R)	−	1
Semi-diameter (SD) (rounded off)	+	16
Observed Altitude (Ho)		47° 43′

	h	m	s		
GHA for	14	50	00	43°	09′
Increment for		8	38	2	10
GHA Sun for	14h	58m	38s	45°	19′
				+360	00
				405	19
Assumed Long. ($a\lambda$) (W)				− 71	19
LHA Sun				334°	00

Declination (Dec.) 05°20.5N

Assumed Latitude (*a*L) 41°N

Tables (see page 88)	Hc	*d*	Z
Tabulation	47° 08′	+52	140°
d Correction	+ 18		
Computed Alt. (Hc)	47° 26′		
Observed Alt. (Ho)	47° 43′		
Intercept (*a*)	17 mi. TOWARD (T)		

(Computed greater away)

Azimuth Angle (Z) N 140° E
True Azimuth (Zn) 140°

You will notice that this sight varies very little from the one given previously. The differences are as follows: DST has been allowed for. It is a morning sight so no conversion from azimuth angle to true azimuth is necessary. Declination and latitude are both north, so we look under 'Declination *same* name as Latitude' in the tables. The *d* correction is plus but this appears clearly in the Tables.

MOON SIGHTS

In my opinion Moon sights are as easy to take and to work out as Sun sights. They can also be extremely useful. For example, when the Moon is waning and visible in the morning sky a simultaneous Sun and Moon sight (one Sun and one Moon sight taken within a few minutes) gives you two position lines and therefore a fix. (See Fig. 22) .

A Moon sight varies very little from a Sun sight. In the Almanac each day you will see a column headed Moon. This gives the GHA and declination for every 10 minutes. Do not interpolate this (or any other) declination in the *Air Almanac* since the time given is the mean for the succeeding period. The declination given for 12:20 is the mean between 12:20 and 12:30, and even if your time is 12:29 you should still take the declination for 12:20.

The GHA increment table for minutes and seconds of GMT is in the same table as that for the Sun, the center column of arc being common to both Sun and Moon. From here the working is exactly the same as for a Sun sight. The difference comes in obtaining the observed from sextant altitude. Index error, dip and refraction naturally remain the same. Semi-diameter is given each day in the Almanac under that of the Sun in the bottom right-hand corner of the a.m. page and this must be added or subtracted depending on whether you take the lower or upper limb (see page 72).

On each page in the Almanac you will see a long, narrow column headed 'Moon's P. in A.' (parallax in altitude). On the left-hand side is the altitude, on the right the correction. The altitude to be used is the sextant altitude and the correction is always added.

An observation of the upper limb of the Moon is worked out below using the page of the Almanac reprinted on page 73.

MOON Upper limb 9 September 1966. DR Lat. 40°55′N. Long. 70°56′W.

	h	m	s
Watch Time (WT)	11	03	34
Watch Error (WE) (s)	+		52
Zone Time (ZT)	11	04	26
Zone description (ZD)	+ 5	00	00
	16	04	26
Daylight Saving Time (DST)	− 1	00	00
Greenwich Mean Time	15^h	04^m	26^s

Sextant altitude (hs)		48° 45.5
Index Error (IE)	−	2
Dip (D)	−	3
Refraction (R)	−	1
Semi-diameter (SD)	−	16
Parallax in Altitude (P. in A.)	+	38
Observed Altitude (Ho)		49° 01.5

	h m s	
GHA for	15 00 00	117° 10′
Increment for	04 26	1 04
GHA Moon for	15h 04m 26s	118° 14′
Assumed Longitude ($a\lambda$) (W)		− 71 14
LHA Moon		47° 00′

Declination (Dec.) 27° 04′N

Assumed Latitude (*a*L) 41°N

Tables *(see page 89)*	Hc	*d*	Z
Tabulation	49° 09′	+32	95
d correction	+ 2		
Computed Alt. (Hc)	49° 11		
Observed Alt. (Ho)	49° 01.5		
Intercept (*a*)	9.5 mi. AWAY (A)		

(Computed greater away)

Azimuth Angle (Z) N 95 W

True Azimuth (Zn) 265°

This sight is plotted in Fig. 22 together with the second Sun sight, thus giving a fix.

PLANET SIGHTS

The planets generally used for navigation are Venus, Jupiter, Mars and Saturn. The *Air Almanac* gives the necessary data for three of these each day. The planets are extremely useful for navigation as they are clearly visible while there is a good horizon in the morning or evening. The GHA and declination of each planet are given and the same increment table for minutes and seconds of GHA is used as for the Sun. There is, of course, no correction for semi-diameter but otherwise the working and plotting is exactly the same as for the Sun.

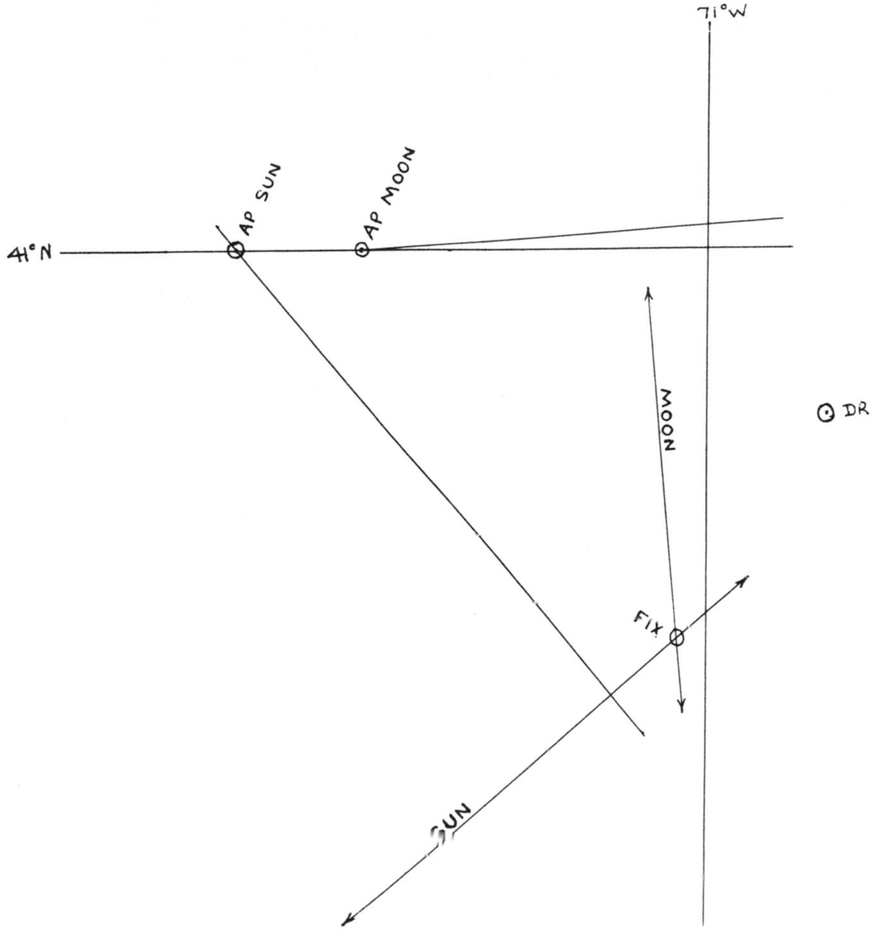

Fig. 22

Pages A58 and A59 of the Almanac give a *Planet Location Diagram* and an explanation of how to use it. It will enable you to identify any unknown planet, but do not expect to do so in a hurry at first. The diagram is complicated and the explanation must be read with care. I find it helpful to think of the times at the top of the diagram as stationary while the diagram, wrapped round a long cylinder, revolves below.

NOON SIGHTS

When taking a noon sight the first step is to find out the time the Sun will cross your meridian. The quickest way, if you are west of Greenwich, is to look down the Sun's GHA column in the Almanac until you find a figure which corresponds to your DR longitude and read off the time against it. For example, on 9 September, (see page 73) with a DR longitude of 23°08.8W, you find 1330, and this is the time (GMT) the Sun crosses the meridian 23°09′W on that particular day. If you are, say 12°E, subtract 12 from 360 (348), and look back up the column for a GHA of about that figure. You will find that the Sun crosses the meridian 12°E just before 1110. Remember that these times are all GMT and must be adjusted accordingly for standard time, etc.

Suppose that you are on the meridian 124°W on 9 September 1966. 124 in the Sun's GHA column lies between 2010 and 2020, and this corrected for a standard time of —8 hours (and DST if necessary), gives a local time of 1210-1220 for your meridian passage. So at 1210 (or earlier if you are not very sure of your time or your longitude), start taking sights. There is no need to time them. You will find that the Sun is still rising slowly, then it will appear to stay at one altitude for a minute or two before beginning to drop. Take the highest altitude as your sight. The working is as follows:

Noon Sight SUN 9 September 1966. DR Lat. 37°42′N.

90°	00′	
—57	23	Observed altitude (Ho)
32°	37	Zenith distance
+05	15.5	Declination N. (Add when same name as Latitude)
37°	52.5N	Latitude

You are 10.5 miles further north than you thought.

Theoretically you might get a wrong answer if you were, say, in Latitude 2° N when the Sun had a declination further north, say, 18°. In fact, the Sun in those circumstances will normally be too high to take a noon sight. We need not therefore consider these cases where declination — altitude = latitude.

As a further example, let us take the case of an imaginary sight in winter in north latitudes:

°	′	
90	00	
—16	38	Observed Altitude.
73	22	Zenith distance.
22	39	S. Declination (subtract when different name from Lat.)
50°	43′	Latitude.

Do not forget that the declination must be taken out for GMT, so that if you are far east or west of Greenwich your noon may be several hours different from Greenwich and therefore the appropriate declination must be found.

'Noon' sights can, of course, be taken for any heavenly body and the approximate GMT of any meridian passage found by inspection of the respective GHA.

Noon sights have three advantages:

1. Accurate time is not necessary.
2. The working out is very simple.
3. There is no plotting on the chart.

STAR SIGHTS

Sidereal Hour Angle

The GHA's of individual stars are not given in the Almanac. A known point in the heavens called the First Point of Aries (denoted by the sign ♈ has been chosen, and the GHA of this point is worked out and given in the Almanac as if it were a heavenly body. The stars for navigational purposes are fixed in their relation to each other and to ♈, so that the angle between the meridian of ♈ and the meridian of a particular star does not change. This angle, measured in a westerly direction from the meridian of ♈, is the Sidereal Hour Angle (SHA).

In Fig. 23:

——→	denotes GHA ♈	The hour angle of the meridian of ♈ measured westward from the Greenwich meridian.
– – →	denotes SHA ★	The hour angle of the star measured westward from the meridian of ♈. This angle does not alter.
—·—·→	denotes GHA ★	The hour angle of the star measured westward from the Greenwich meridian.

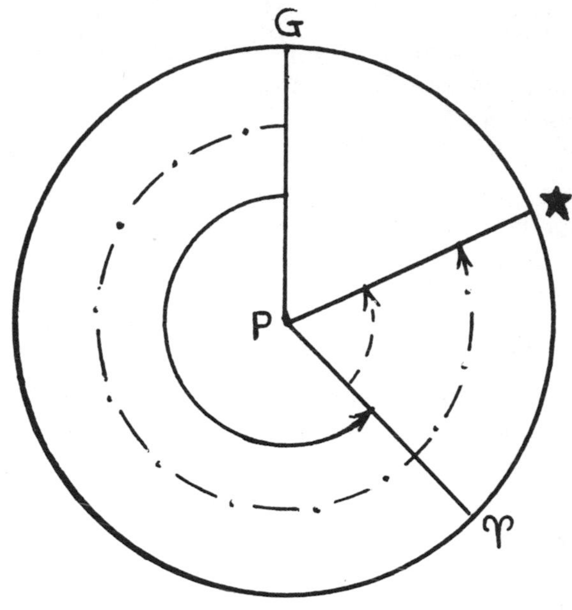

Fig. 23

This last angle *always* equals the sum of the other two hour angles. GHA ♈ + SHA ★ = GHA ★

Let us look at an example. On 1 October 1966, at 03:20 (page 74)—

GHA ♈ at 0320	59° 25′		This figure is constant and is found inside the front cover of the Almanac and also on the flap following the buff colored daily pages, (see page 76).
SHA *Sirius*	259 05		
GHA *Sirius*	318° 30′		

The table for the increment to be added to GHA ♈ for the minutes and seconds of GMT is the same as that for the Sun. Having gotten GHA ♈, LHA ♈ is obtained in the usual way.

Declination

Declinations of stars do not vary during the year and are given inside the front cover and on the flap of the Almanac. It follows then that provided the declination of the star is less than 30° (either North or South) the sight can be worked out from Volumes II and III of H.O. 249, same as for a planet. Volume I, however, is designed for star sights and should normally be used.

H.O. 249, Vol. I, Selected Stars

The object of this volume is to provide for every hour, anywhere, the necessary data for seven selected stars. If only three stars are to be observed, those marked with an asterisk (*) should be used, as they will give the best cuts. Star sights can only be taken, save in very exceptional conditions, at morning or evening twilight when both the stars and the horizon are visible. It is not easy to pick up a star in the sextant, so that it is advisable to know the approximate altitude and azimuth of the star you are going to take. The altitude is set on the sextant and by taking a rough bearing it is possible to find the star in the sextant almost on the horizon immediately. Suppose that you wish to take star sights on the evening of 1 October 1966, and your position is approximately 30°N, 76°W. Pages A62-A67 of the Almanac give the times of sunrise and sunset and morning and evening civil twilight tabulated at an interval of three days. It is usually accurate enough to use the times given for the nearest tabular date. Civil twilight lasts from sunset until the sun is 6° below the horizon and from when the sun is 6° below the horizon until sunrise. Its duration can, therefore, be found from the length of time between the times given in the sunset (or sunrise) and evening (or morning) civil twilight tables, respectively. In good weather sights can be taken after (or in the morning, before) civil twilight when both the stars and the horizon are visible. The darker limit of civil twilight (sun 6° below the horizon) should be used as the mid-time for your observations.

Now on 1 October 1966 (nearest tabulated date 2 October), at 30°N sunset is at 1745 and twilight at 1809 on the Greenwich Meridian, (see page 78), so that civil twilight lasts for only twenty-four minutes. Your meridian, however, is 76°W, so twilight will be later by 76° or 5 hours 4 minutes (the table for conversion of arc into time is on page A79 in the Almanac) (see page 79), so you should plan to make your observations around 2313 GMT, or 1813 zone time (ZD + 5). GHA ♈ at 2313 GMT is 358° to take the nearest whole degree, and subtracting 76°W longitude we arrive at LHA ♈ 282. In H.O. 249, Vol. I, turn to the page headed 'LAT 30°N', find LHA 282, (see page 83), and against it you will find the tabulated altitudes and the true azimuths, Zn (instead of azimuth angles, Z) of the seven stars best suited for that time and place. They are *Deneb, Altair, Nunki, Antares, Arcturus, Alkaid* and *Kochab. Deneb, Nuki* and *Arcturus* are starred and should be used if only three observations are made.

Let us take another example: Early in the morning of 9 September 1966, you are approaching the coast of Europe, DR 49°N, 08°W. Sunrise on the Greenwich meridian is at 0525 and twilight begins at 0451 GMT, so you have about 34 minutes of twilight. You are 8° west of Greenwich, so twilight will be (8 x 4) 32 minutes later on your meridian, or about 0523. GHA ♈ at 0520 is 67° 48′, and to obtain LHA we must subtract 7° 48′ of longitude to get a round figure of LHA ♈ 60°. H.O. 249, Vol. I, (see page 84), LAT. 49°N, LHA ♈ 60° tells us which are the best stars. If you can, observe at least five of them, but for this example we assume that you take sights of the three that are starred: *Dubhe, Rigel* and *Alpheratz.*

The accurate taking of star sights is a great art. Some can see the horizon or the stars long before others, some can read their own sextant and not be blinded by the light, while others prefer to have their sextant read for them. Only practice can help you on these points.

At least five sights of each star should be taken and then averaged. To obtain observed altitude from sextant altitude, index error, dip and refraction must be applied, but not, of course, semi-diameter. Your three sights will look something like this:

9 September 1966. DR Lat. 48°58′ N. Long 08°10′W.

	Dubhe	*Rigel*	*Alpheratz*
	h m s	h m s	h m s
GMT	05 20 03	05 16 48	05 13 26
GHA ♈	67°49′	65°18′	65°18′
Increment	01	1 42	52
GHA ♈	67°50′	67°00′	66°10′
a λ (W)	– 7 50	– 8 00	– 8 10
LHA ♈	60°00′	59°00′	58°00′

Tables (H.O. 249, Vol. I)

Computed Alt. (Hc)	35°44′	30°20′	43°04′
True Azimuth (Zn)	034°	158°	266°

These altitudes can now be compared with the observed altitudes and the intercepts plotted on the chart. Remember that these are true azimuths, and that no further conversion is necessary.

There is one further correction, for precession and nutation. 'Table 5—Precession and Nutation Correction' at the end of H.O. 249, Vol. I, (see page 85), should be entered with the year, latitude and LHA ♈ and a correction may be found which is applied, not to the individual sight, but to the fix on the chart (or if only one star has been used, to the position line) . A distance and bearing is given by which the whole fix is to be moved. In this case, 1966, Lat. 50° (nearest) , and LHA ♈ 060°, we see that the fix must be moved 1 nautical mile in true bearing 070°.

If you wish to take sights of stars with declinations higher than 30° and which are not in Vol. I, you must either use other tables such as H.O. 214, (see page 68) , or revert to the cosine-haversine method.

Stars can be identified from a navigational star chart, a star finder and identifier (see page 66), a star globe, or from the Sky Diagrams at the end of the *Air Almanac*. These latter if studied carefully will be found excellent either for finding the position of a star which you wish to locate or for finding what you *have* taken if you do not know.

POLE STAR SIGHTS

As has already been said (page 29) , if *Polaris* were directly over the North Pole, its observed altitude would be your latitude. But, unfortunately, *Polaris* can bear 2°6′ east or west of north, so a correction must be applied.

When you observe *Polaris* notice the time. The nearest minute or so will do. Then look up GHA ♈ in the Almanac for that time and find LHA. Your sight was at 0445 GMT 1 October 1966:

GHA ♈ for 0440	79°28′	Sextant Altitude (hs)		36°10′
DR Long. (E)	+16 13	Index error (IE)	—	2
LHA	95°41′	Dip (D)	—	3
		Refraction (R)	—	1
		Observed Alt. (Ho)		36°04′

Turn to the last page of the Almanac where you will find a table headed '*POLARIS* (Pole Star) TABLE 1966,' (see page 80) . Look against LHA ♈ 95°41′ and you will find — 22.

Subtract this from your observed altitude, and the answer, 35°42′ is your latitude.

You will have noticed that there is a great similarity in all the sights that have been discussed. The following is a summary of the directions given:

SUN, MOON, PLANETS: enter the Almanac with GMT. Enter the tables with LHA, declination and assumed latitude. Compare the result with the observed altitude.

STARS: enter the Almanac with GMT. Enter H.O. 249, Vol. I, with LHA ♈ and assumed latitude, or H.O. 249, Vols. II and III, with LHA star, declination and assumed latitude. Compare answer with observed altitude.

NOON SIGHT: zenith distance ± declination = latitude.

POLARIS: enter the Almanac with the approximate GMT. Enter *Polaris* table with LHA ♈. Computed altitude = latitude.

Two points remain to be stressed:

First, with modern methods sights are easy to work out. Time, thanks to radio, presents no problem, but *the accuracy of your position line depends on the accuracy of your sight.* You may have a poor horizon, a rough sea, spray coming over the sextant or clouds tending to hide the Sun. Any of these will increase your difficulties and no book can help you to overcome them. The only answer is practice and then more practice until you gain skill and confidence.

Secondly, do not imagine that celestial navigation is only of use if you are crossing the Atlantic. It can be of the greatest value in coastal work. After a night's drifting, for instance, a sight which gives you a position line parallel to your course may enable you to alter it some time before landmarks can be identified, and this alteration may save you several hours sailing against the current.

Accurate DR is, of course, the basis of good yacht na tion but errors creep in, and to have your DR supported corrected by sights in which you have confidence may sav you hours of worry or annoyance. When you can trust your sights you will find that you can use them frequently to good purpose and I hope this book will not only enable you to get your position lines but also encourage you to plunge deeper into the fascinating study of celestial navigation.

SECTION THREE: NOTES

Sextants

When you look through the eye-piece of your sextant you see part of a rectangular frame. The left-hand side of this is plain glass through which you can see the horizon, while the right-hand side is a mirror. This reflects the light from another mirror at the top of the sextant which is fixed to and swings with the main arm. The bottom of this arm swings along a scale. This scale is calibrated in degrees, and the minutes are read off a wheel by which the small adjustments are made. In older sextants the minutes are read off the degree scale with the help of a vernier.

Cover the top mirror with one of the tinted shades, then face the Sun and look at the horizon (or the garden fence!) through the eye-piece. Swing the arm very slowly until the Sun appears, then with the small adjuster wheel 'move' the Sun until it rests on the horizon. This is your sight. Read off your sextant to the nearest minute. The result is your sextant altitude.

The sextant is a delicate instrument and must be treated carefully. It is liable to various errors of which the chief are:

1. Sextant error

This is the basic error of the sextant and should be marked by the manufacturer on the lid of the sextant box.

2. Index error

This is a variable error and should be checked frequently in the following manner: set the sextant roughly at zero and look at the Sun. You will see two suns, bring them so that their edges are just touching and read the sextant. Then reverse the suns and read the sextant again. You will find that one reading will be on the ordinary scale (on), and one on the minus side (off). Subtract the smaller from the greater, halve it, and the result is the index error, to be added if the greater number were off and subtracted if it were on. You can test the accuracy of your reading by the fact that the two figures added together should equal four times the semi-diameter of the Sun. Your figures may read like this:

34′.5 on, 30′.25 off.

$$\begin{array}{r} 34.5 \\ 30.25 \\ \hline 4.25 \div 2 = 2'.13 \text{ on.} \end{array}$$

Taken to the nearest minute, the index error is minus 2′ and this must always be subtracted from your sextant altitude.

Another way of getting index error if you have a really good hard horizon is to set your sextant at zero, look at the horizon and make it into one straight line and then look at your sextant reading. It *should* but will not be zero. Sextants should be corrected if the index error is over 3′.

An imaginary index error of —2′ has been taken for the examples throughout this book.

3. Side error

When you are checking the index error the Suns should appear exactly one above the other. If they are very much out of alignment the sextant should be corrected.

There are other sextant errors. These, together with methods of correction, can be found in Bowditch.

Star Identification

Star Finder and Identifier (H.O. 2102-D)

This is one of the most commonly used star finders. It is published by the U. S. Navy Oceanographic Office. It consists of a white opaque disc base with a projection of all the navigational stars which are listed in the Almanacs. One side is for the Northern Hemisphere, and the other, the Southern Hemisphere. In addition, it contains a series of transparent templates, at 10° intervals of latitude, with each template having a family of altitude and azimuth curves.

To use the star finder, you must determine LHA ♈, then select the template which includes your DR latitude, and place it on the base plate of the appropriate hemisphere (N or S). By rotating the template so that the arrow is over LHA ♈, the approximate altitudes and azimuths of the stars visible above the horizon are indicated by the curves.

The star finder can be used to help you locate predetermined stars, as well as identify a star which you shot but were unable to identify.

An additional template is also included which is used to plot additional bodies, such as the Sun, Moon, Planets or additional stars.

The *Star Finder and Identifier* is very often commonly referred to, erroneously, as the *Rude Star Finder.*

Sky Diagrams

At the back of the *Air Almanac* are 17 pages devoted to morning and evening sky diagrams for different months, latitudes and hours. They are extremely useful but no explanation will be given here as one cannot hope to improve on the explanation given in the Almanac. I do not, however, advise readers to bother about them until they are fairly conversant with the principles of celestial navigation.

Spherical Triangles

For those interested, the basic formulae for solving the spherical triangle are given below:

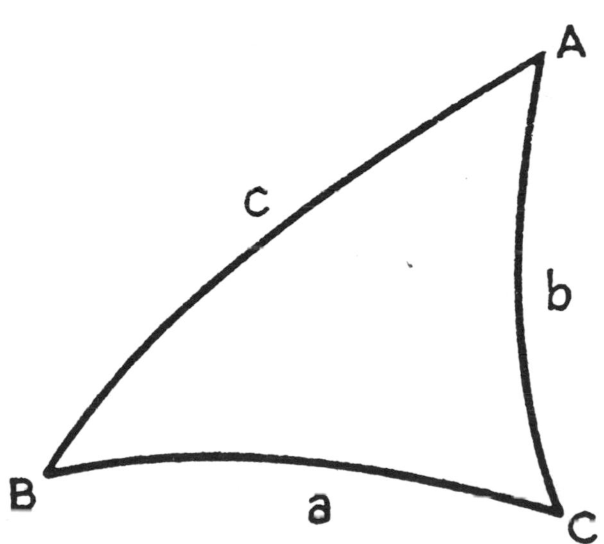

$$\text{Cos a} = \text{Cos b cos c} + \text{sin b sin c cos A}$$

$$\text{Cos C} = \frac{\text{Cos c} - \text{cos a cos b}}{\text{Sin a sin b}}$$

Additional Tables

Tables of Computed Altitude and Azimuth (H.O. 214)

Each of the nine volumes of these tables covers ten degrees of latitude for declinations 0—89°. They should be used with the *Nautical Almanac* if greater accuracy is required than is obtainable from H.O. 249. Altitude is given to 0·1′ and azimuth to 0·1°. These tables are more complicated than H.O. 249 and therefore there is more possibility of making an error. The explanations are simple and very lucid, but if you are going to use them, work out a sight or two on dry land first. Don't use them for the first time when tired and urgently needing an accurate position.

Astronomical Navigation Tables (H.O. 218)

These are as easy to use and a good deal easier to read, as well as being smaller, than H.O.249. They are compiled on the same principle as H.O.249 but there is one main difference. Refraction has been allowed for in these tables and when using them *no* refraction correction should be applied to the sextant altitude. The star section at the beginning of each volume must be entered with LHA star and not LHA ♈

Final Note

South Latitudes

Only north latitudes have been considered in this book. The only difference, however, for working out a sight in south latitudes is that the azimuth is taken from the south. An azimuth angle of 45° will be S 45°E or S 45° W depending on circumstances.

APPENDICES

APPENDIX A – NAVIGATION ABBREVIATIONS and SYMBOLS

APPENDIX B – EXCERPTS from THE AIR ALMANAC

APPENDIX C – EXCERPTS from H.O. 249, Vol. I

APPENDIX D – EXCERPTS from H.O. 249, Vol. II

APPENDIX E – EXCERPTS from H.O. 249, Vol. III

APPENDIX A

NAVIGATION ABBREVIATIONS and SYMBOLS

The following abbreviations and symbols are only those which are used throughout this book.

Altitude	H, h
Computed	Hc
Observed	Ho
Sextant	hs
Difference (intercept)	*a*
Aries	♈
Assumed latitude	*a*L
Assumed longitude	*a*λ
Assumed position	AP
Away (altitude difference)	A
Azimuth (true)	Zn
Azimuth angle	Z, Az
Daylight Saving Time	DST
Declination	Dec.
Interpolation value	*d*
Degrees	°
Dead reckoning	DR
Dip	D
East	E
Fast (watch error)	f
Geographical position	GP
Greenwich Hour Angle	GHA
Greenwich Mean Time	GMT
Hours	h
Index error	IE
Intercept (altitude difference)	*a*
Latitude	Lat., L

Local Hour Angle	LHA
Longitude	Long., λ
Minutes (arc)	′
Minutes (time)	m
North	N
Observed altitude	Ho
Parallax in Altitude	P in A
Refraction	R
Seconds (time)	s
Semi-diameter	SD
Sextant altitude	hs
Sidereal Hour Angle	SHA
Slow (watch error)	s
Toward (altitude difference)	T
Watch error	WE
Watch time	WT
West	W
Zone description	ZD
Zone time	ZT

APPENDIX B — AIR ALMANAC

GMT	⊙ SUN GHA	SUN Dec.	ARIES GHA ♈	VENUS — 3.3 GHA	VENUS Dec.	JUPITER — 1.5 GHA	JUPITER Dec.	SATURN 0.9 GHA	SATURN Dec.	◑ MOON GHA	MOON Dec.
h m	° ′	° ′	° ′	° ′	° ′	° ′	° ′	° ′	° ′	° ′	° ′
00 00	180 35.9	N 5 34.5	347 35.6	195 04	N12 33	228 36	N20 57	349 17	S 3 25	260 59	N26 36
10	183 06.0	34.3	350 06.0	197 34		231 07		351 48		263 23	36
20	185 36.0	34.2	352 36.4	200 04		233 37		354 18		265 47	37
30	188 06.0	· 34.0	355 06.8	202 34	· ·	236 07	· ·	356 49	· ·	268 11	· 37
40	190 36.1	33.8	357 37.2	205 04		238 38		359 19		270 36	38
50	193 06.1	33.7	0 07.6	207 34		241 08		1 50		273 00	38
01 00	195 36.2	N 5 33.5	2 38.0	210 04	N12 32	243 38	N20 57	4 20	S 3 25	275 24	N26 39
10	198 06.2	33.4	5 08.5	212 34		246 09		6 50		277 48	39
20	200 36.2	33.2	7 38.9	215 03		248 39		9 21		280 12	40
30	203 06.3	· 33.1	10 09.3	217 33	· ·	251 09	· ·	11 51	· ·	282 37	· 40
40	205 36.3	32.9	12 39.7	220 03		253 40		14 22		285 01	41
50	208 06.3	32.7	15 10.1	222 33		256 10		16 52		287 25	41
02 00	210 36.4	N 5 32.6	17 40.5	225 03	N12 31	258 40	N20 57	19 23	S 3 25	289 49	N26 42
10	213 06.4	32.4	20 10.9	227 33		261 11		21 53		292 13	42
20	215 36.4	32.3	22 41.3	230 03		263 41		24 24		294 37	43
30	218 06.5	· 32.1	25 11.7	232 33	· ·	266 11	· ·	26 54	· ·	297 02	· 43
40	220 36.5	32.0	27 42.1	235 03		268 42		29 24		299 26	44
50	223 06.5	31.8	30 12.6	237 33		271 12		31 55		301 50	44
03 00	225 36.6	N 5 31.6	32 43.0	240 03	N12 30	273 42	N20 57	34 25	S 3 25	304 14	N26 44
10	228 06.6	31.5	35 13.4	242 32		276 13		36 56		306 38	45
20	230 36.7	31.3	37 43.8	245 02		278 43		39 26		309 03	45
30	233 06.7	· 31.2	40 14.2	247 32	· ·	281 13	· ·	41 57	· ·	311 27	· 46
40	235 36.7	31.0	42 44.6	250 02		283 44		44 27		313 51	46
50	238 06.8	30.9	45 15.0	252 32		286 14		46 57		316 15	46
04 00	240 36.8	N 5 30.7	47 45.4	255 02	N12 29	288 44	N20 57	49 28	S 3 26	318 39	N26 47
10	243 06.8	30.6	50 15.8	257 32		291 15		51 58		321 03	47
20	245 36.9	30.4	52 46.3	260 02		293 45		54 29		323 28	48
30	248 06.9	· 30.2	55 16.7	262 32	· ·	296 15	· ·	56 59	· ·	325 52	· 48
40	250 36.9	30.1	57 47.1	265 02		298 46		59 30		328 16	49
50	253 07.0	29.9	60 17.5	267 32		301 16		62 00		330 40	49
05 00	255 37.0	N 5 29.8	62 47.9	270 02	N12 28	303 46	N20 56	64 31	S 3 26	333 04	N26 49
10	258 07.0	29.6	65 18.3	272 31		306 17		67 01		335 28	50
20	260 37.1	29.5	67 48.7	275 01		308 47		69 31		337 52	50
30	263 07.1	· 29.3	70 19.1	277 31	· ·	311 17	· ·	72 02	· ·	340 17	· 50
40	265 37.2	29.1	72 49.5	280 01		313 48		74 32		342 41	51
50	268 07.2	29.0	75 20.0	282 31		316 18		77 03		345 05	51
06 00	270 37.2	N 5 28.8	77 50.4	285 01	N12 27	318 48	N20 56	79 33	S 3 26	347 29	N26 51
10	273 07.3	28.7	80 20.8	287 31		321 19		82 04		349 53	52
20	275 37.3	28.5	82 51.2	290 01		323 49		84 34		352 17	52
30	278 07.3	· 28.4	85 21.6	292 31	· ·	326 19	· ·	87 05	· ·	354 41	· 52
40	280 37.4	28.2	87 52.0	295 01		328 50		89 35		357 06	53
50	283 07.4	28.0	90 22.4	297 31		331 20		92 05		359 30	53
07 00	285 37.4	N 5 27.9	92 52.8	300 01	N12 26	333 50	N20 56	94 36	S 3 26	1 54	N26 53
10	288 07.5	27.7	95 23.2	302 30		336 21		97 06		4 18	54
20	290 37.5	27.6	97 53.6	305 00		338 51		99 37		6 42	54
30	293 07.6	· 27.4	100 24.1	307 30	· ·	341 21	· ·	102 07	· ·	9 06	· 54
40	295 37.6	27.3	102 54.5	310 00		343 52		104 38		11 30	55
50	298 07.6	27.1	105 24.9	312 30		346 22		107 08		13 54	55
08 00	300 37.7	N 5 26.9	107 55.3	315 00	N12 25	348 52	N20 56	109 38	S 3 26	16 18	N26 55
10	303 07.7	26.8	110 25.7	317 30		351 23		112 09		18 43	56
20	305 37.7	26.6	112 56.1	320 00		353 53		114 39		21 07	56
30	308 07.8	· 26.5	115 26.5	322 30	· ·	356 23	· ·	117 10	· ·	23 31	· 56
40	310 37.8	26.3	117 56.9	325 00		358 54		119 40		25 55	57
50	313 07.8	26.2	120 27.3	327 30		1 24		122 11		28 19	57
09 00	315 37.9	N 5 26.0	122 57.8	329 59	N12 24	3 54	N20 56	124 41	S 3 26	30 43	N26 57
10	318 07.9	25.8	125 28.2	332 29		6 25		127 12		33 07	57
20	320 37.9	25.7	127 58.6	334 59		8 55		129 42		35 31	58
30	323 08.0	· 25.5	130 29.0	337 29	· ·	11 25	· ·	132 12	· ·	37 55	· 58
40	325 38.0	25.4	132 59.4	339 59		13 56		134 43		40 19	58
50	328 08.1	25.2	135 29.8	342 29		16 26		137 13		42 44	58
10 00	330 38.1	N 5 25.1	138 00.2	344 59	N12 23	18 56	N20 56	139 44	S 3 26	45 08	N26 59
10	333 08.1	24.9	140 30.6	347 29		21 27		142 14		47 32	59
20	335 38.2	24.7	143 01.0	349 59		23 57		144 45		49 56	59
30	338 08.2	· 24.6	145 31.5	352 29	· ·	26 27	· ·	147 15	· ·	52 20	26 59
40	340 38.2	24.4	148 01.9	354 59		28 58		149 46		54 44	27 00
50	343 08.3	24.3	150 32.3	357 28		31 28		152 16		57 08	00
11 00	345 38.3	N 5 24.1	153 02.7	359 58	N12 22	33 58	N20 56	154 46	S 3 26	59 32	N27 00
10	348 08.3	24.0	155 33.1	2 28		36 29		157 17		61 56	00
20	350 38.4	23.8	158 03.5	4 58		38 59		159 47		64 20	00
30	353 08.4	· 23.6	160 33.9	7 28	· ·	41 29	· ·	162 18	· ·	66 44	· 01
40	355 38.4	23.5	163 04.3	9 58		44 00		164 48		69 08	01
50	358 08.5	23.3	165 34.7	12 28		46 30		167 19		71 33	01

Lat.	Moon-rise	Diff.
N		
°	h m	m
72	□	*
70	□	*
68	□	*
66	□	*
64	□	*
62	20 33	33
60	21 20	32
58	21 50	32
56	22 13	32
54	22 32	32
52	22 48	32
50	23 02	32
45	23 30	31
40	23 52	31
35	24 10	31
30	24 26	31
20	24 53	31
10	00 17	28
0	00 38	29
10	00 59	30
20	01 22	30
30	01 49	31
35	02 04	31
40	02 22	32
45	02 44	32
50	03 12	33
52	03 26	33
54	03 42	34
56	04 01	35
58	04 24	36
60	04 54	37
S		

Moon's P. in A.

° Alt.	- + Corr.	° Alt.	- + Corr.
0		54	
	58		33
6		55	
	57		32
12		57	
	56		31
16		58	
	55		30
19		59	
	54		29
22		60	
	53		28
24		61	
	52		27
27		62	
	51		26
29		63	
	50		25
31		64	
	49		24
33		66	
	48		23
34		67	
	47		22
36		68	
	46		21
38		69	
	45		20
39		70	
	44		19
41		71	
	43		18
42		72	
	42		17
44		73	
	41		16
45		74	
	40		15
46		75	
	39		14
48		76	
	38		13
49		77	
	37		12
50		78	
	36		11
52		79	
	35		10
53		80	
	34		
54			
	33		
55			

⊙Sun SD 15′.9
Moon SD 16′
Age 24ᵈ

504 GREENWICH P. M. 1966 SEPTEMBER 9 (FRIDAY)

GMT	☉ SUN GHA	☉ SUN Dec.	ARIES GHA ♈	VENUS —3.3 GHA	VENUS Dec.	JUPITER —1.5 GHA	JUPITER Dec.	SATURN 0.9 GHA	SATURN Dec.	◑ MOON GHA	◑ MOON Dec.
h m	° ′	° ′	° ′	° ′	° ′	° ′	° ′	° ′	° ′	° ′	° ′
12 00	0 38.5	N 5 23.2	168 05.1	14 58	N12 21	49 00	N20 56	169 49	S 3 26	73 57	N27 01
10	3 08.6	23.0	170 35.6	17 28		51 31		172 19		76 21	01
20	5 38.6	22.9	173 06.0	19 58		54 01		174 50		78 45	02
30	8 08.6	· 22.7	175 36.4	22 28	· ·	56 31	· ·	177 20	· ·	81 09	· 02
40	10 38.7	22.5	178 06.8	24 58		59 02		179 51		83 33	02
50	13 08.7	22.4	180 37.2	27 27		61 32		182 21		85 57	02
13 00	15 38.7	N 5 22.2	183 07.6	29 57	N12 19	64 02	N20 56	184 52	S 3 26	88 21	N27 02
10	18 08.8	22.1	185 38.0	32 27		66 33		187 22		90 45	02
20	20 38.8	21.9	188 08.4	34 57		69 03		189 53		93 09	03
30	23 08.8	· 21.8	190 38.8	37 27	· ·	71 33	· ·	192 23	· ·	95 33	· 03
40	25 38.9	21.6	193 09.3	39 57		74 03		194 53		97 57	03
50	28 08.9	21.4	195 39.7	42 27		76 34		197 24		100 21	03
14 00	30 38.9	N 5 21.3	198 10.1	44 57	N12 18	79 04	N20 56	199 54	S 3 26	102 45	N27 03
10	33 09.0	21.1	200 40.5	47 27		81 34		202 25		105 09	03
20	35 39.0	21.0	203 10.9	49 57		84 05		204 55		107 33	03
30	38 09.1	· 20.8	205 41.3	52 27	· ·	86 35	· ·	207 26	· ·	109 57	· 04
40	40 39.1	20.6	208 11.7	54 57		89 05		209 56		112 21	04
50	43 09.1	20.5	210 42.1	57 26		91 36		212 26		114 45	04
15 00	45 39.2	N 5 20.3	213 12.5	59 56	N12 17	94 06	N20 56	214 57	S 3 26	117 10	N27 04
10	48 09.2	20.2	215 42.9	62 26		96 36		217 27		119 34	04
20	50 39.2	20.0	218 13.4	64 56		99 07		219 58		121 58	04
30	53 09.3	· 19.9	220 43.8	67 26	· ·	101 37	· ·	222 28	· ·	124 22	· 04
40	55 39.3	19.7	223 14.2	69 56		104 07		224 59		126 46	04
50	58 09.3	19.5	225 44.6	72 26		106 38		227 29		129 10	04
16 00	60 39.4	N 5 19.4	228 15.0	74 56	N12 16	109 08	N20 56	230 00	S 3 26	131 34	N27 04
10	63 09.4	19.2	230 45.4	77 26		111 38		232 30		133 58	04
20	65 39.5	19.1	233 15.8	79 56		114 09		235 00		136 22	04
30	68 09.5	· 18.9	235 46.2	82 26	· ·	116 39	· ·	237 31	· ·	138 46	· 05
40	70 39.5	18.8	238 16.6	84 56		119 09		240 01		141 10	05
50	73 09.6	18.6	240 47.1	87 25		121 40		242 32		143 34	05
17 00	75 39.6	N 5 18.4	243 17.5	89 55	N12 15	124 10	N20 56	245 02	S 3 27	145 58	N27 05
10	78 09.6	18.3	245 47.9	92 25		126 40		247 33		148 22	05
20	80 39.7	18.1	248 18.3	94 55		129 11		250 03		150 46	05
30	83 09.7	· 18.0	250 48.7	97 25	· ·	131 41	· ·	252 34	· ·	153 10	· 05
40	85 39.7	17.8	253 19.1	99 55		134 11		255 04		155 34	05
50	88 09.8	17.7	255 49.5	102 25		136 42		257 34		157 58	05
18 00	90 39.8	N 5 17.5	258 19.9	104 55	N12 14	139 12	N20 55	260 05	S 3 27	160 22	N27 05
10	93 09.8	17.3	260 50.3	107 25		141 42		262 35		162 46	05
20	95 39.9	17.2	263 20.8	109 55		144 13		265 06		165 10	05
30	98 09.9	· 17.0	265 51.2	112 25	· ·	146 43	· ·	267 36	· ·	167 34	· 05
40	100 40.0	16.9	268 21.6	114 54		149 13		270 07		169 58	05
50	103 10.0	16.7	270 52.0	117 24		151 44		272 37		172 22	05
19 00	105 40.0	N 5 16.6	273 22.4	119 54	N12 13	154 14	N20 55	275 07	S 3 27	174 46	N27 05
10	108 10.1	16.4	275 52.8	122 24		156 44		277 38		177 10	05
20	110 40.1	16.2	278 23.2	124 54		159 15		280 08		179 34	05
30	113 10.1	· 16.1	280 53.6	127 24	· ·	161 45	· ·	282 39	· ·	181 58	· 05
40	115 40.2	15.9	283 24.0	129 54		164 15		285 09		184 22	05
50	118 10.2	15.8	285 54.4	132 24		166 46		287 40		186 46	05
20 00	120 40.2	N 5 15.6	288 24.9	134 54	N12 12	169 16	N20 55	290 10	S 3 27	189 10	N27 05
10	123 10.3	15.5	290 55.3	137 24		171 46		292 41		191 34	05
20	125 40.3	15.3	293 25.7	139 54		174 17		295 11		193 58	05
30	128 10.4	· 15.1	295 56.1	142 24	· ·	176 47	· ·	297 41	· ·	196 22	· 05
40	130 40.4	15.0	298 26.5	144 53		179 17		300 12		198 46	05
50	133 10.4	14.8	300 56.9	147 23		181 48		302 42		201 10	04
21 00	135 40.5	N 5 14.7	303 27.3	149 53	N12 11	184 18	N20 55	305 13	S 3 27	203 34	N27 04
10	138 10.5	14.5	305 57.7	152 23		186 48		307 43		205 58	04
20	140 40.5	14.4	308 28.1	154 53		189 19		310 14		208 22	04
30	143 10.6	· 14.2	310 58.6	157 23	· ·	191 49	· ·	312 44	· ·	210 46	· 04
40	145 40.6	14.0	313 29.0	159 53		194 19		315 15		213 10	04
50	148 10.6	13.9	315 59.4	162 23		196 50		317 45		215 34	04
22 00	150 40.7	N 5 13.7	318 29.8	164 53	N12 10	199 20	N20 55	320 15	S 3 27	217 58	N27 04
10	153 10.7	13.6	321 00.2	167 23		201 50		322 46		220 22	04
20	155 40.7	13.4	323 30.6	169 53		204 21		325 16		222 46	04
30	158 10.8	· 13.3	326 01.0	172 23	· ·	206 51	· ·	327 47	· ·	225 09	· 04
40	160 40.8	13.1	328 31.4	174 52		209 21		330 17		227 33	03
50	163 10.9	12.9	331 01.8	177 22		211 52		332 48		229 57	03
23 00	165 40.9	N 5 12.8	333 32.3	179 52	N12 09	214 22	N20 55	335 18	S 3 27	232 21	N27 03
10	168 10.9	12.6	336 02.7	182 22		216 52		337 48		234 45	03
20	170 41.0	12.5	338 33.1	184 52		219 23		340 19		237 09	03
30	173 11.0	· 12.3	341 03.5	187 22	· ·	221 53	· ·	342 49	· ·	239 33	· 03
40	175 41.0	12.1	343 33.9	189 52		224 23		345 20		241 57	03
50	178 11.1	12.0	346 04.3	192 22		226 54		347 50		244 21	02

Lat.	Moonset	Diff.
N		
°	h m	m
72	□	*
70	□	*
68	□	*
66	□	*
64	□	*
62	18 11	30
60	17 24	30
58	16 53	31
56	16 30	31
54	16 11	31
52	15 56	31
50	15 42	31
45	15 14	30
40	14 52	30
35	14 33	30
30	14 18	30
20	13 51	30
10	13 28	30
0	13 06	30
10	12 45	29
20	12 21	29
30	11 55	29
35	11 39	29
40	11 20	29
45	10 58	28
50	10 30	28
52	10 16	27
54	10 00	27
56	09 41	26
58	09 18	25
60	08 47	24
S		

Moon's P. in A.

° Alt.	+ Corr. ′	° Alt.	+ Corr. ′
0	58	54	33
9	57	56	32
14	56	57	31
17	55	58	30
20	54	59	29
23	53	60	28
25	52	61	27
27	51	62	26
29	50	64	25
31	49	65	24
33	48	66	23
35	47	67	22
37	46	68	21
38	45	69	20
40	44	70	19
41	43	71	18
43	42	72	17
44	41	73	16
45	40	74	15
47	39	75	14
48	38	76	13
49	37	77	12
51	36	78	11
52	35	79	10
53	34	80	
54	33		
56			

☉Sun SD 15′.9
Moon SD 16′
Age 24ᵈ

GREENWICH A. M. 1966 OCTOBER 1 (SATURDAY)

GMT	⊙ SUN GHA	⊙ SUN Dec.	ARIES GHA ♈	VENUS — 3.4 GHA	VENUS Dec.	JUPITER — 1.6 GHA	JUPITER Dec.	SATURN 0.8 GHA	SATURN Dec.	○ MOON GHA	○ MOON Dec.
h m	° ′	° ′	° ′	° ′	° ′	° ′	° ′	° ′	° ′	° ′	° ′
00 00	182 30.7	S 2 55.5	9 16.6	191 10	N 2 23	246 31	N20 16	12 31	S 4 06	348 30	N 6 25
10	185 00.8	55.6	11 47.0	193 40		249 01		15 02		350 56	27
20	187 30.8	55.8	14 17.4	196 10		251 32		17 32		353 22	29
30	190 00.8	· 55.9	16 47.9	198 40 ·	·	254 02 ·	·	20 02 ·	·	355 47	· 31
40	192 30.9	56.1	19 18.3	201 10		256 33		22 33		358 13	34
50	195 00.9	56.3	21 48.7	203 40		259 03		25 03		0 39	36
01 00	197 31.0	S 2 56.4	24 19.1	206 10	N 2 22	261 33	N20 16	27 34	S 4 06	3 05	N 6 38
10	200 01.0	56.6	26 49.5	208 40		264 04		30 04		5 31	40
20	202 31.0	56.8	29 19.9	211 09		266 34		32 35		7 57	43
30	205 01.1	· 56.9	31 50.3	213 39 ·	·	269 04 ·	·	35 05 ·	·	10 23	· 45
40	207 31.1	57.1	34 20.7	216 09		271 35		37 35		12 49	47
50	210 01.1	57.2	36 51.1	218 39		274 05		40 06		15 15	49
02 00	212 31.2	S 2 57.4	39 21.5	221 09	N 2 21	276 35	N20 16	42 36	S 4 06	17 41	N 6 52
10	215 01.2	57.6	41 52.0	223 39		279 06		45 07		20 06	54
20	217 31.2	57.7	44 22.4	226 09		281 36		47 37		22 32	56
30	220 01.3	· 57.9	46 52.8	228 39 ·	·	284 06 ·	·	50 08 ·	·	24 58	6 58
40	222 31.3	58.1	49 23.2	231 09		286 37		52 38		27 24	7 00
50	225 01.3	58.2	51 53.6	233 39		289 07		55 09		29 50	03
03 00	227 31.4	S 2 58.4	54 24.0	236 09	N 2 20	291 37	N20 15	57 39	S 4 06	32 16	N 7 05
10	230 01.4	58.5	56 54.4	238 39		294 08		60 09		34 42	07
20	232 31.4	58.7	59 24.8	241 09		296 38		62 40		37 08	09
30	235 01.5	· 58.9	61 55.2	243 39 ·	·	299 08 ·	·	65 10 ·	·	39 34	· 12
40	237 31.5	59.0	64 25.7	246 09		301 39		67 41		41 59	14
50	240 01.5	59.2	66 56.1	248 38		304 09		70 11		44 25	16
04 00	242 31.6	S 2 59.3	69 26.5	251 08	N 2 18	306 40	N20 15	72 42	S 4 07	46 51	N 7 18
10	245 01.6	59.5	71 56.9	253 38		309 10		75 12		49 17	20
20	247 31.6	59.7	74 27.3	256 08		311 40		77 43		51 43	23
30	250 01.7	2 59.8	76 57.7	258 38 ·	·	314 11 ·	·	80 13 ·	·	54 09	· 25
40	252 31.7	3 00.0	79 28.1	261 08		316 41		82 43		56 35	27
50	255 01.7	00.2	81 58.5	263 38		319 11		85 14		59 01	29
05 00	257 31.8	S 3 00.3	84 28.9	266 08	N 2 17	321 42	N20 15	87 44	S 4 07	61 27	N 7 32
10	260 01.8	00.5	86 59.3	268 38		324 12		90 15		63 52	34
20	262 31.8	00.6	89 29.8	271 08		326 42		92 45		66 18	36
30	265 01.9	· 00.8	92 00.2	273 38 ·	·	329 13 ·	·	95 16 ·	·	68 44	· 38
40	267 31.9	01.0	94 30.6	276 08		331 43		97 46		71 10	40
50	270 01.9	01.1	97 01.0	278 38		334 13		100 16		73 36	43
06 00	272 32.0	S 3 01.3	99 31.4	281 08	N 2 16	336 44	N20 15	102 47	S 4 07	76 02	N 7 45
10	275 02.0	01.4	102 01.8	283 38		339 14		105 17		78 28	47
20	277 32.0	01.6	104 32.2	286 07		341 44		107 48		80 53	49
30	280 02.1	· 01.8	107 02.6	288 37 ·	·	344 15 ·	·	110 18 ·	·	83 19	· 51
40	282 32.1	01.9	109 33.0	291 07		346 45		112 49		85 45	54
50	285 02.1	02.1	112 03.5	293 37		349 15		115 19		88 11	56
07 00	287 32.2	S 3 02.3	114 33.9	296 07	N 2 15	351 46	N20 15	117 50	S 4 07	90 37	N 7 58
10	290 02.2	02.4	117 04.3	298 37		354 16		120 20		93 03	8 00
20	292 32.2	02.6	119 34.7	301 07		356 47		122 50		95 29	02
30	295 02.3	· 02.7	122 05.1	303 37 ·	·	359 17 ·	·	125 21 ·	·	97 55	· 05
40	297 32.3	02.9	124 35.5	306 07		1 47		127 51		100 20	07
50	300 02.3	03.1	127 05.9	308 37		4 18		130 22		102 46	09
08 00	302 32.4	S 3 03.2	129 36.3	311 07	N 2 13	6 48	N20 15	132 52	S 4 07	105 12	N 8 11
10	305 02.4	03.4	132 06.7	313 37		9 18		135 23		107 38	14
20	307 32.4	03.6	134 37.2	316 07		11 49		137 53		110 04	16
30	310 02.5	· 03.7	137 07.6	318 37 ·	·	14 19 ·	·	140 24 ·	·	112 30	· 18
40	312 32.5	03.9	139 38.0	321 07		16 49		142 54		114 56	20
50	315 02.5	04.0	142 08.4	323 36		19 20		145 24		117 21	22
09 00	317 32.6	S 3 04.2	144 38.8	326 06	N 2 12	21 50	N20 15	147 55	S 4 07	119 47	N 8 25
10	320 02.6	04.4	147 09.2	328 36		24 20		150 25		122 13	27
20	322 32.6	04.5	149 39.6	331 06		26 51		152 56		124 39	29
30	325 02.7	· 04.7	152 10.0	333 36 ·	·	29 21 ·	·	155 26 ·	·	127 05	· 31
40	327 32.7	04.8	154 40.4	336 06		31 51		157 57		129 31	33
50	330 02.7	05.0	157 10.8	338 36		34 22		160 27		131 56	36
10 00	332 32.8	S 3 05.2	159 41.3	341 06	N 2 11	36 52	N20 15	162 57	S 4 07	134 22	N 8 38
10	335 02.8	05.3	162 11.7	343 36		39 22		165 28		136 48	40
20	337 32.9	05.5	164 42.1	346 06		41 53		167 58		139 14	42
30	340 02.9	· 05.7	167 12.5	348 36 ·	·	44 23 ·	·	170 29 ·	·	141 40	· 44
40	342 32.9	05.8	169 42.9	351 06		46 53		172 59		144 06	47
50	345 03.0	06.0	172 13.3	353 36		49 24		175 30		146 31	49
11 00	347 33.0	S 3 06.1	174 43.7	356 06	N 2 10	51 54	N20 15	178 00	S 4 07	148 57	N 8 51
10	350 03.0	06.3	177 14.1	358 36		54 24		180 31		151 23	53
20	352 33.1	06.5	179 44.5	1 05		56 55		183 01		153 49	55
30	355 03.1	· 06.6	182 15.0	3 35 ·	·	59 25 ·	·	185 31 ·	·	156 15	8 57
40	357 33.1	06.8	184 45.4	6 05		61 56		188 02		158 41	9 00
50	0 03.2	06.9	187 15.8	8 35		64 26		190 32		161 06	02

Lat.	Moon-rise	Diff.
N		
°	h m	m
72	17 02	−20
70	17 17	−14
68	17 30	−10
66	17 40	−06
64	17 49	−03
62	17 56	−01
60	18 03	+01
58	18 09	03
56	18 14	04
54	18 18	05
52	18 23	06
50	18 26	07
45	18 35	10
40	18 42	12
35	18 48	13
30	18 53	15
20	19 02	17
10	19 11	19
0	19 19	21
10	19 26	23
20	19 35	25
30	19 45	28
35	19 50	29
40	19 56	31
45	20 04	33
50	20 13	35
52	20 17	36
54	20 22	38
56	20 27	39
58	20 33	41
60	20 39	43
S		

Moon's P. in A.

° Alt.	+ Corr. ′	° Alt.	+ Corr. ′
0		57	
	54		29
10		58	
	53		28
15		59	
	52		27
18		60	
	51		26
21		62	
	50		25
24		63	
	49		24
26		64	
	48		23
29		65	
	47		22
31		66	
	46		21
33		67	
	45		20
35		68	
	44		19
36		70	
	43		18
38		71	
	42		17
40		72	
	41		16
41		73	
	40		15
43		74	
	39		14
44		75	
	38		13
46		76	
	37		12
47		77	
	36		11
49		78	
	35		10
50		79	
	34		
51			
	33		
53			
	32		
54			
	31		
55			
	30		
57			
	29		
58			

⊙Sun SD 16′.0

Moon SD 15′

Age 16ᵈ

548 GREENWICH P. M. 1966 OCTOBER 1 (SATURDAY)

GMT	☉ SUN GHA	☉ SUN Dec.	ARIES GHA ♈	VENUS — 3.4 GHA	VENUS — 3.4 Dec.	JUPITER — 1.6 GHA	JUPITER — 1.6 Dec.	SATURN 0.8 GHA	SATURN 0.8 Dec.	☽ MOON GHA	☽ MOON Dec.
h m	° ′	° ′	° ′	° ′	° ′	° ′	° ′	° ′	° ′	° ′	° ′
12 00	2 33.2	S 3 07.1	189 46.2	11 05	N 2 08	66 56	N20 15	193 03	S 4 07	163 32	N 9 04
10	5 03.2	07.3	192 16.6	13 35		69 27		195 33		165 58	06
20	7 33.3	07.4	194 47.0	16 05		71 57		198 04		168 24	08
30	10 03.3	· 07.6	197 17.4	18 35 ·	·	74 27 ·	·	200 34 ·	·	170 50	· 11
40	12 33.3	07.8	199 47.8	21 05		76 58		203 04		173 16	13
50	15 03.4	07.9	202 18.2	23 35		79 28		205 35		175 41	15
13 00	17 33.4	S 3 08.1	204 48.7	26 05	N 2 07	81 58	N20 15	208 05	S 4 07	178 07	N 9 17
10	20 03.4	08.2	207 19.1	28 35		84 29		210 36		180 33	19
20	22 33.5	08.4	209 49.5	31 05		86 59		213 06		182 59	21
30	25 03.5	· 08.6	212 19.9	33 35 ·	·	89 29 ·	·	215 37 ·	·	185 25	· 24
40	27 33.5	08.7	214 50.3	36 05		92 00		218 07		187 51	26
50	30 03.6	08.9	217 20.7	38 35		94 30		220 38		190 16	28
14 00	32 33.6	S 3 09.0	219 51.1	41 05	N 2 06	97 00	N20 15	223 08	S 4 07	192 42	N 9 30
10	35 03.6	09.2	222 21.5	43 34		99 31		225 38		195 08	32
20	37 33.7	09.4	224 51.9	46 04		102 01		228 09		197 34	35
30	40 03.7	· 09.5	227 22.3	48 34 ·	·	104 31 ·	·	230 39 ·	·	200 00	· 37
40	42 33.7	09.7	229 52.8	51 04		107 02		233 10		202 25	39
50	45 03.8	09.9	232 23.2	53 34		109 32		235 40		204 51	41
15 00	47 33.8	S 3 10.0	234 53.6	56 04	N 2 05	112 03	N20 15	238 11	S 4 07	207 17	N 9 43
10	50 03.8	10.2	237 24.0	58 34		114 33		240 41		209 43	45
20	52 33.9	10.3	239 54.4	61 04		117 03		243 12		212 09	48
30	55 03.9	· 10.5	242 24.8	63 34 ·	·	119 34 ·	·	245 42 ·	·	214 35	· 50
40	57 33.9	10.7	244 55.2	66 04		122 04		248 12		217 00	52
50	60 04.0	10.8	247 25.6	68 34		124 34		250 43		219 26	54
16 00	62 34.0	S 3 11.0	249 56.0	71 04	N 2 03	127 05	N20 15	253 13	S 4 07	221 52	N 9 56
10	65 04.0	11.1	252 26.5	73 34		129 35		255 44		224 18	9 58
20	67 34.1	11.3	254 56.9	/6 04		132 05		258 14		226 44	10 01
30	70 04.1	· 11.5	257 27.3	78 34 ·	·	134 36 ·	·	260 45 ·	·	229 09	· 03
40	72 34.1	11.6	259 57.7	81 03		137 06		263 15		231 35	05
50	75 04.2	11.8	262 28.1	83 33		139 36		265 45		234 01	07
17 00	77 34.2	S 3 12.0	264 58.5	86 03	N 2 02	142 07	N20 14	268 16	S 4 07	236 27	N10 09
10	80 04.2	12.1	267 28.9	88 33		144 37		270 46		238 52	11
20	82 34.3	12.3	269 59.3	91 03		147 07		273 17		241 18	14
30	85 04.3	· 12.4	272 29.7	93 33 ·	·	149 38 ·	·	275 47 ·	·	243 44	· 16
40	87 34.3	12.6	275 00.2	96 03		152 08		278 18		246 10	18
50	90 04.4	12.8	277 30.6	98 33		154 38		280 48		248 36	20
18 00	92 34.4	S 3 12.9	280 01.0	101 03	N 2 01	157 09	N20 14	283 19	S 4 08	251 01	N10 22
10	95 04.4	13.1	282 31.4	103 33		159 39		285 49		253 27	24
20	97 34.5	13.2	285 01.8	106 03		162 10		288 19		255 53	27
30	100 04.5	· 13.4	287 32.2	108 33 ·	·	164 40 ·	·	290 50 ·	·	258 19	· 29
40	102 34.5	13.6	290 02.6	111 03		167 10		293 20		260 45	31
50	105 04.6	13.7	292 33.0	113 33		169 41		295 51		263 10	33
19 00	107 34.6	S 3 13.9	295 03.4	116 03	N 2 00	172 11	N20 14	298 21	S 4 08	265 36	N10 35
10	110 04.6	14.1	297 33.8	118 32		174 41		300 52		268 02	37
20	112 34.7	14.2	300 04.3	121 02		177 12		303 22		270 28	39
30	115 04.7	· 14.4	302 34.7	123 32 ·	·	179 42 ·	·	305 53 ·	·	272 53	· 42
40	117 34.7	14.5	305 05.1	126 02		182 12		308 23		275 19	44
50	120 04.8	14.7	307 35.5	128 32		184 43		310 53		277 45	46
20 00	122 34.8	S 3 14.9	310 05.9	131 02	N 1 58	187 13	N20 14	313 24	S 4 08	280 11	N10 48
10	125 04.8	15.0	312 36.3	133 32		189 43		315 54		282 36	50
20	127 34.9	15.2	315 06.7	136 02		192 14		318 25		285 02	52
30	130 04.9	· 15.4	317 37.1	138 32 ·	·	194 44 ·	·	320 55 ·	·	287 28	· 54
40	132 34.9	15.5	320 07.5	141 02		197 14		323 26		289 54	57
50	135 05.0	15.7	322 38.0	143 32		199 45		325 56		292 20	10 59
21 00	137 35.0	S 3 15.8	325 08.4	146 02	N 1 57	202 15	N20 14	328 26	S 4 08	294 45	N11 01
10	140 05.0	16.0	327 38.8	148 32		204 45		330 57		297 11	03
20	142 35.1	16.2	330 09.2	151 02		207 16		333 27		299 37	05
30	145 05.1	· 16.3	332 39.6	153 32 ·	·	209 46 ·	·	335 58 ·	·	302 03	· 07
40	147 35.1	16.5	335 10.0	156 01		212 17		338 28		304 28	09
50	150 05.2	16.6	337 40.4	158 31		214 47		340 59		306 54	12
22 00	152 35.2	S 3 16.8	340 10.8	161 01	N 1 56	217 17	N20 14	343 29	S 4 08	309 20	N11 14
10	155 05.2	17.0	342 41.2	163 31		219 48		346 00		311 46	16
20	157 35.3	17.1	345 11.6	166 01		222 18		348 30		314 11	18
30	160 05.3	· 17.3	347 42.1	168 31 ·	·	224 48 ·	·	351 00 ·	·	316 37	· 20
40	162 35.3	17.5	350 12.5	171 01		227 19		353 31		319 03	22
50	165 05.4	17.6	352 42.9	173 31		229 49		356 01		321 28	24
23 00	167 35.4	S 3 17.8	355 13.3	176 01	N 1 55	232 19	N20 14	358 32	S 4 08	323 54	N11 26
10	170 05.4	17.9	357 43.7	178 31		234 50		1 02		326 20	29
20	172 35.5	18.1	0 14.1	181 01		237 20		3 33		328 46	31
30	175 05.5	· 18.3	2 44.5	183 31 ·	·	239 50 ·	·	6 03 ·	·	331 11	· 33
40	177 35.5	18.4	5 14.9	186 01		242 21		8 33		333 37	35
50	180 05.6	18.6	7 45.3	188 31		244 51		11 04		336 03	37

Lat.	Moon-set	Diff.
N		
°	h m	m
72	08 46	64
70	08 33	57
68	08 23	52
66	08 14	49
64	08 07	46
62	08 01	44
60	07 56	42
58	07 51	40
56	07 47	38
54	07 43	37
52	07 40	36
50	07 37	35
45	07 30	32
40	07 25	30
35	07 20	29
30	07 16	27
20	07 09	25
10	07 03	23
0	06 57	21
10	06 51	19
20	06 45	17
30	06 38	14
35	06 34	13
40	06 30	11
45	06 24	09
50	06 18	07
52	06 15	06
54	06 12	05
56	06 09	04
58	06 05	02
60	06 01	01
S		

Moon's P. in A.

° Alt.	′ + Corr.	° Alt.	′ + Corr.
0		56	
	55		30
3		57	
	54		29
11		58	
	53		28
15		59	
	52		27
19		60	
	51		26
22		62	
	50		25
24		63	
	49		24
27		64	
	48		23
29		65	
	47		22
31		66	
	46		21
33		67	
	45		20
35		69	
	44		19
37		70	
	43		18
38		71	
	42		17
40		72	
	41		16
42		73	
	40		15
43		74	
	39		14
45		75	
	38		13
46		76	
	37		12
48		77	
	36		11
49		78	
	35		10
50		79	
	34		
52			
	33		
53			
	32		
54			
	31		
56			
	30		
57			

☉Sun SD 16′·0

Moon SD 15′

Age 17ᵈ

STARS, SEPT.—DEC., 1966

No.	Name		Mag.	S.H.A.	Dec.
				° ′	° ′
7*	*Acamar*		3·1	315 45	S.40 26
5*	*Achernar*		0·6	335 52	S.57 24
30*	*Acrux*		1·1	173 50	S.62 55
19	*Adhara*	†	1·6	255 40	S.28 55
10*	*Aldebaran*	†	1·1	291 30	N.16 27
32*	*Alioth*		1·7	166 52	N.56 08
34*	*Alkaid*		1·9	153 27	N.49 29
55	*Al Na'ir*		2·2	28 28	S.47 08
15	*Alnilam*	†	1·8	276 22	S. 1 13
25*	*Alphard*	†	2·2	218 31	S. 8 31
41*	*Alphecca*	†	2·3	126 41	N.26 50
1*	*Alpheratz*	†	2·2	358 20	N.28 55
51*	*Altair*	†	0·9	62 43	N. 8 47
2	*Ankaa*		2·4	353 50	S.42 29
42*	*Antares*	†	1·2	113 10	S.26 22
37*	*Arcturus*	†	0·2	146 28	N.19 21
43	*Atria*		1·9	108 44	S.68 58
22	*Avior*		1·7	234 33	S.59 24
13	*Bellatrix*	†	1·7	279 10	N. 6 19
16*	*Betelgeuse*	†	0·1–1·2	271 39	N. 7 24
17*	*Canopus*		−0·9	264 12	S.52 40
12*	*Capella*		0·2	281 26	N.45 58
53*	*Deneb*		1·3	49 56	N.45 10
28*	*Denebola*	†	2·2	183 10	N.14 46
4*	*Diphda*	†	2·2	349 31	S.18 10
27*	*Dubhe*		2·0	194 35	N.61 56
14	*Elnath*	†	1·8	278 57	N.28 35
47	*Eltanin*		2·4	91 03	N.51 30
54*	*Enif*	†	2·5	34 22	N. 9 43
56*	*Fomalhaut*	†	1·3	16 03	S.29 48
31	*Gacrux*		1·6	172 41	S.56 55
29*	*Gienah*	†	2·8	176 29	S.17 21
35	*Hadar*		0·9	149 39	S.60 13
6*	*Hamal*	†	2·2	328 40	N.23 18
48	*Kaus Aust.*		2·0	84 31	S.34 24
40*	*Kochab*		2·2	137 19	N.74 17
57	*Markab*	†	2·6	14 13	N.15 02
8*	*Menkar*	†	2·8	314 52	N. 3 58
36	*Menkent*		2·3	148 50	S.36 12
24*	*Miaplacidus*		1·8	221 48	S.69 34
9*	*Mirfak*		1·9	309 31	N.49 45
50*	*Nunki*	†	2·1	76 42	S.26 21
52*	*Peacock*		2·1	54 15	S.56 51
21*	*Pollux*	†	1·2	244 11	N.28 07
20*	*Procyon*	†	0·5	245 37	N. 5 19
46*	*Rasalhague*	†	2·1	96 39	N.12 35
26*	*Regulus*	†	1·3	208 21	N.12 08
11*	*Rigel*	†	0·3	281 46	S. 8 14
38*	*Rigil Kent.*		0·1	140 41	S.60 42
44	*Sabik*	†	2·6	102 53	S.15 41
3*	*Schedar*		2·5	350 21	N.56 21
45*	*Shaula*		1·7	97 10	S.37 05
18*	*Sirius*	†	−1·6	259 05	S.16 40
33*	*Spica*	†	1·2	159 09	S.10 59
23*	*Suhail*		2·2	223 19	S.43 18
49*	*Vega*		0·1	81 03	N.38 45
39	*Zuben'ubi*	†	2·9	137 45	S.15 54

INTERPOLATION OF G.H.A.

Increment to be added for intervals of G.M.T. to G.H.A. of: Sun, Aries (♈) and planets ; Moon

SUN, etc.		MOON	SUN, etc.		MOON	SUN, etc.		MOON
m s	° ′	m s	m s	° ′	m s	m s	° ′	m s
00 00	0 00	00 00	03 17	0 50	03 25	06 37	1 40	06 52
01	0 01	00 02	21	0 51	03 29	41	1 41	06 56
05	0 02	00 06	25	0 52	03 33	45	1 42	07 00
09	0 03	00 10	29	0 53	03 37	49	1 43	07 04
13	0 04	00 14	33	0 54	03 41	53	1 44	07 08
17	0 05	00 18	37	0 55	03 45	06 57	1 45	07 13
21	0 06	00 22	41	0 56	03 49	07 01	1 46	07 17
25	0 07	00 26	45	0 57	03 54	05	1 47	07 21
29	0 08	00 31	49	0 58	03 58	09	1 48	07 25
33	0 09	00 35	53	0 59	04 02	13	1 49	07 29
37	0 10	00 39	03 57	1 00	04 06	17	1 50	07 33
41	0 11	00 43	04 01	1 01	04 10	21	1 51	07 37
45	0 12	00 47	05	1 02	04 14	25	1 52	07 42
49	0 13	00 51	09	1 03	04 19	29	1 53	07 46
53	0 14	00 55	13	1 04	04 23	33	1 54	07 50
00 57	0 15	01 00	17	1 05	04 27	37	1 55	07 54
01 01	0 16	01 04	21	1 06	04 31	41	1 56	07 58
05	0 17	01 08	25	1 07	04 35	45	1 57	08 02
09	0 18	01 12	29	1 08	04 39	49	1 58	08 06
13	0 19	01 16	33	1 09	04 43	53	1 59	08 11
17	0 20	01 20	37	1 10	04 48	07 57	2 00	08 15
21	0 21	01 24	41	1 11	04 52	08 01	2 01	08 19
25	0 22	01 29	45	1 12	04 56	05	2 02	08 23
29	0 23	01 33	49	1 13	05 00	09	2 03	08 27
33	0 24	01 37	53	1 14	05 04	13	2 04	08 31
37	0 25	01 41	04 57	1 15	05 08	17	2 05	08 35
41	0 26	01 45	05 01	1 16	05 12	21	2 06	08 40
45	0 27	01 49	05	1 17	05 17	25	2 07	08 44
49	0 28	01 53	09	1 18	05 21	29	2 08	08 48
53	0 29	01 58	13	1 19	05 25	33	2 09	08 52
01 57	0 30	02 02	17	1 20	05 29	37	2 10	08 56
02 01	0 31	02 06	21	1 21	05 33	41	2 11	09 00
05	0 32	02 10	25	1 22	05 37	45	2 12	09 04
09	0 33	02 14	29	1 23	05 41	49	2 13	09 09
13	0 34	02 18	33	1 24	05 46	53	2 14	09 13
17	0 35	02 22	37	1 25	05 50	08 57	2 15	09 17
21	0 36	02 27	41	1 26	05 54	09 01	2 16	09 21
25	0 37	02 31	45	1 27	05 58	05	2 17	09 25
29	0 38	02 35	49	1 28	06 02	09	2 18	09 29
33	0 39	02 39	53	1 29	06 06	13	2 19	09 33
37	0 40	02 43	05 57	1 30	06 10	17	2 20	09 38
41	0 41	02 47	06 01	1 31	06 15	21	2 21	09 42
45	0 42	02 51	05	1 32	06 19	25	2 22	09 46
49	0 43	02 56	09	1 33	06 23	29	2 23	09 50
53	0 44	03 00	13	1 34	06 27	33	2 24	09 54
02 57	0 45	03 04	17	1 35	06 31	37	2 25	09 58
03 01	0 46	03 08	21	1 36	06 35	41	2 26	10 00
05	0 47	03 12	25	1 37	06 39	45	2 27	
09	0 48	03 16	29	1 38	06 44	49	2 28	
13	0 49	03 20	33	1 39	06 48	53	2 29	
17	0 50	03 25	37	1 40	06 52	09 57	2 30	
03 21		03 29	06 41		06 56	10 00		

* Stars used in H.O. 249 (A.P. 3270) Vol. 1.
† Stars that may be used with Vols. 2 and 3.

A62

SUNRISE

Lat.	Aug.	September										October				Lat.
	30	2	5	8	11	14	17	20	23	26	29	2	5	8	11	
°	h m	h m	h m	h m	h m	h m	h m	h m	h m	h m	h m	h m	h m	h m	h m	°
N 72	03 49	04 04	04 18	04 33	04 47	05 00	05 14	05 27	05 41	05 54	06 08	06 21	06 35	06 49	07 03	N 72
70	04 04	17	30	42	04 54	06	18	30	42	54	06	18	30	42	06 54	70
68	17	28	39	50	05 01	11	22	32	43	54	04	15	26	36	47	68
66	27	37	47	04 56	06	15	25	34	44	53	03	12	22	32	41	66
64	35	44	53	05 02	10	19	27	36	44	53	02	10	19	28	36	64
62	43	51	04 59	06	14	22	30	37	45	53	06 01	08	16	24	32	62
N 60	04 49	04 56	05 03	05 10	05 17	05 24	05 31	05 38	05 45	05 52	05 59	06 07	06 14	06 21	06 28	N 60
58	54	05 01	07	14	20	27	33	39	46	52	59	05	12	18	25	58
56	04 59	05	11	17	23	28	34	40	46	52	58	04	10	16	22	56
54	05 04	09	14	20	25	30	36	41	46	52	57	03	08	14	19	54
52	07	12	17	22	27	32	37	42	47	52	57	02	07	12	17	52
N 50	05 11	05 16	05 20	05 24	05 29	05 33	05 38	05 42	05 47	05 51	05 56	06 01	06 05	06 10	06 15	N 50
45	19	22	26	29	33	37	40	44	48	51	55	05 58	02	06	10	45
40	25	28	31	34	37	39	42	45	48	51	54	57	06 00	06 03	06	40
35	31	33	35	37	39	42	44	46	48	50	53	55	05 57	05 59	06 02	35
30	35	37	39	40	42	44	45	47	48	50	52	54	55	57	05 59	30
N 20	05 44	05 44	05 45	05 46	05 46	05 47	05 48	05 48	05 49	05 50	05 50	05 51	05 52	05 52	05 53	N 20
N 10	51	51	51	50	50	50	50	49	49	49	49	49	48	48	48	N 10
0	05 57	05 56	05 55	54	53	52	51	50	49	48	47	46	45	44	44	0
S 10	06 04	06 02	06 00	05 59	05 57	55	53	51	49	47	46	44	42	40	39	S 10
20	11	08	05	06 03	06 00	05 57	55	52	49	47	44	41	38	36	33	20
S 30	06 18	06 15	06 11	06 08	06 04	06 00	05 57	05 53	05 49	05 45	05 42	05 38	05 34	05 31	05 27	S 30
35	23	19	14	10	06	02	57	53	49	44	40	36	32	28	24	35
40	27	23	18	13	08	03	05 59	54	49	44	39	34	29	24	19	40
45	33	28	22	17	11	05	06 00	54	48	43	37	31	26	20	15	45
50	40	33	27	21	14	08	01	54	48	41	35	28	22	15	09	50
S 52	06 43	06 36	06 29	06 22	06 16	06 09	06 02	05 55	05 48	05 41	05 34	05 27	05 20	05 13	05 06	S 52
54	46	39	32	25	17	10	02	55	47	40	33	25	18	10	03	54
56	50	42	35	27	19	11	03	55	47	39	31	23	15	08	05 00	56
58	54	46	38	29	21	12	04	55	47	38	30	21	13	05	04 56	58
S 60	06 59	06 50	06 41	06 32	06 23	06 14	06 05	05 56	05 47	05 37	05 28	05 19	05 10	05 01	04 52	S 60

SUNSET

Lat.	Aug.	September										October				Lat.
	30	2	5	8	11	14	17	20	23	26	29	2	5	8	11	
°	h m	h m	h m	h m	h m	h m	h m	h m	h m	h m	h m	h m	h m	h m	h m	°
N 72	20 10	19 53	19 36	19 20	19 04	18 49	18 33	18 18	18 02	17 47	17 31	17 16	17 01	16 45	16 30	N 72
70	19 54	40	26	11	18 57	43	29	15	01	47	34	20	06	52	38	70
68	42	30	17	19 04	51	38	26	13	01	48	35	23	10	16 58	45	68
66	33	21	09	18 58	46	35	23	11	18 00	48	37	26	14	17 03	51	66
64	25	14	19 03	53	42	31	21	10	17 59	49	38	28	17	07	16 56	64
62	17	08	18 58	48	38	29	19	09	59	49	39	30	20	10	17 01	62
N 60	19 11	19 02	18 53	18 44	18 35	18 26	18 17	18 08	17 58	17 49	17 40	17 31	17 22	17 13	17 04	N 60
58	06	18 57	49	41	32	24	15	07	58	50	41	33	25	16	08	58
56	19 01	53	45	38	30	22	14	06	58	50	42	34	26	18	11	56
54	18 57	49	42	35	27	20	12	05	58	50	43	35	28	21	14	54
52	53	46	39	32	25	18	11	04	57	50	43	36	29	23	16	52
N 50	18 49	18 43	18 37	18 30	18 24	18 17	18 10	18 04	17 57	17 51	17 44	17 38	17 31	17 25	17 18	N 50
45	42	36	31	25	20	14	08	03	57	51	46	40	34	29	23	45
40	36	31	26	21	16	11	06	01	56	51	47	42	37	32	27	40
35	30	26	22	18	13	09	05	01	56	52	48	43	39	35	31	35
30	25	22	18	15	11	07	04	18 00	56	52	49	45	41	38	34	30
N 20	18 17	18 15	18 12	18 10	18 07	18 04	18 01	17 59	17 56	17 53	17 50	17 48	17 45	17 43	17 40	N 20
N 10	10	09	07	05	03	18 01	18 00	58	56	54	52	50	49	47	45	N 10
0	18 04	18 03	18 02	18 01	18 00	17 59	17 58	57	56	55	54	53	52	51	50	0
S 10	17 58	17 58	17 57	17 57	17 57	57	56	56	56	56	55	55	55	17 55	17 55	S 10
20	51	52	52	53	54	54	55	55	56	57	17 57	17 58	17 59	18 00	18 01	20
S 30	17 44	17 45	17 47	17 48	17 50	17 52	17 53	17 55	17 56	17 58	18 00	18 02	18 03	18 05	18 07	S 30
35	39	42	44	46	48	50	52	55	57	17 59	01	04	06	09	11	35
40	35	37	40	43	46	49	51	54	57	18 00	03	06	09	12	15	40
45	29	33	36	40	43	47	50	54	58	01	05	09	12	16	20	45
50	23	27	31	36	40	45	49	54	58	03	07	12	17	21	26	50
S 52	17 20	17 24	17 29	17 34	17 39	17 44	17 49	17 53	17 58	18 03	18 08	18 13	18 18	18 24	18 29	S 52
54	16	21	27	32	37	43	48	53	59	04	10	15	21	26	32	54
56	13	18	24	30	36	41	47	53	59	05	11	17	23	29	35	56
58	09	15	21	27	34	40	46	53	17 59	06	12	19	25	32	39	58
S 60	17 04	17 11	17 18	17 25	17 32	17 39	17 46	17 53	18 00	18 07	18 14	18 21	18 28	18 36	18 43	S 60

MORNING CIVIL TWILIGHT

Lat.	Aug.	September										October				Lat.
	30	2	5	8	11	14	17	20	23	26	29	2	5	8	11	
°	h m	h m	h m	h m	h m	h m	h m	h m	h m	h m	h m	h m	h m	h m	h m	°
N 72	02 12	02 35	02 55	03 14	03 31	03 47	04 03	04 18	04 32	04 46	05 00	05 14	05 27	05 41	05 54	N 72
70	02 45	03 02	03 18	33	03 48	04 01	15	28	40	53	05	17	29	41	52	70
68	03 08	22	36	03 48	04 01	13	24	36	47	04 58	09	19	30	41	51	68
66	27	38	03 50	04 01	11	22	32	42	52	05 02	12	21	31	40	50	66
64	41	03 51	04 01	11	20	30	39	48	04 57	06	14	23	32	40	49	64
62	03 53	04 02	11	19	28	36	44	52	05 01	09	16	24	32	40	48	62
N 60	04 03	04 13	04 19	04 27	04 34	04 42	04 49	04 57	05 04	05 11	05 18	05 25	05 33	05 40	05 47	N 60
58	12	19	26	33	40	47	53	05 00	07	13	20	26	33	39	46	58
56	19	26	32	38	45	51	04 57	03	09	15	21	27	33	39	45	56
54	26	32	37	43	49	54	05 00	06	11	17	22	27	33	38	44	54
52	32	37	42	47	53	04 58	03	08	13	18	23	28	33	38	43	52
N 50	04 37	04 42	04 47	04 51	04 56	05 01	05 05	05 10	05 15	05 19	05 24	05 29	05 33	05 38	05 42	N 50
45	48	04 52	04 56	05 00	05 03	07	11	15	18	22	26	29	33	37	40	45
40	04 57	05 00	05 03	06	09	12	15	18	21	24	27	30	33	35	38	40
35	05 05	07	09	12	14	16	18	21	23	25	27	30	32	34	37	35
30	11	13	15	16	18	20	21	23	25	26	28	30	31	33	35	30
N 20	05 21	05 22	05 23	05 24	05 24	05 25	05 26	05 26	05 27	05 28	05 28	05 29	05 30	05 30	05 31	N 20
N 10	30	29	29	29	29	29	29	28	28	28	28	28	27	27	27	N 10
0	36	35	35	34	33	32	31	30	29	28	27	25	25	24	23	0
S 10	43	41	39	38	36	34	32	30	28	26	25	23	21	19	18	S 10
20	48	46	43	41	38	35	33	30	27	24	22	19	16	14	11	20
S 30	05 54	05 51	05 47	05 44	05 40	05 36	05 33	05 29	05 25	05 21	05 18	05 14	05 10	05 07	05 03	S 30
35	05 57	53	49	45	41	37	32	28	24	19	15	11	06	05 02	04 58	35
40	06 00	55	51	46	41	36	32	27	22	17	12	07	05 02	04 57	52	40
45	03	05 58	53	47	42	36	31	25	19	13	07	05 02	04 56	50	44	45
50	07	06 01	55	48	42	35	29	22	16	09	02	04 56	49	42	36	50
S 52	06 09	06 02	05 56	05 49	05 42	05 35	05 28	05 21	05 14	05 07	05 00	04 53	04 45	04 38	04 31	S 52
54	10	03	56	49	42	35	27	20	12	05	04 57	49	42	34	26	54
56	12	05	57	50	42	34	26	18	10	05 02	54	46	38	29	21	56
58	14	06	58	50	42	33	25	16	08	04 59	50	42	33	24	15	58
S 60	06 16	06 08	05 59	05 50	05 42	05 33	05 24	05 14	05 05	04 56	04 46	04 37	04 27	04 18	04 08	S 60

EVENING CIVIL TWILIGHT

Lat.	Aug.	September										October				Lat.
	30	2	5	8	11	14	17	20	23	26	29	2	5	8	11	
°	h m	h m	h m	h m	h m	h m	h m	h m	h m	h m	h m	h m	h m	h m	h m	°
N 72	21 44	21 20	20 58	20 38	20 19	20 01	19 44	19 27	19 10	18 54	18 38	18 23	18 08	17 53	17 38	N 72
70	21 12	20 54	36	19	20 03	19 47	32	17	19 03	48	34	20	07	53	40	70
68	20 50	34	19	20 04	19 50	36	23	09	18 56	43	31	18	06	53	41	68
66	32	19	20 06	19 53	40	27	15	19 03	51	39	28	16	05	54	43	66
64	18	20 06	19 55	43	32	20	09	18 58	47	36	25	15	04	54	44	64
62	20 07	19 56	45	35	24	14	19 04	53	43	33	23	14	04	54	45	62
N 60	19 57	19 47	19 37	19 28	19 18	19 08	18 59	18 50	18 40	18 31	18 22	18 13	18 04	17 55	17 46	N 60
58	48	39	30	22	13	19 04	55	46	38	29	20	12	03	55	47	58
56	41	32	24	16	08	18 59	51	43	35	27	19	11	03	56	48	56
54	34	26	19	11	03	56	48	40	33	25	18	10	03	56	49	54
52	28	21	14	07	19 00	52	45	38	31	24	17	10	03	56	50	52
N 50	19 23	19 16	19 10	19 03	18 56	18 50	18 43	18 36	18 29	18 23	18 16	18 10	18 03	17 57	17 51	N 50
45	12	19 07	19 01	18 55	49	43	38	32	26	20	15	09	03	58	53	45
40	19 03	18 59	18 54	49	44	39	34	28	23	18	13	09	04	17 59	55	40
35	18 56	52	48	43	39	35	30	26	21	17	13	09	04	18 00	56	35
30	50	46	43	39	35	31	27	24	20	16	12	09	05	02	17 58	30
N 20	18 40	18 37	18 34	18 32	18 29	18 26	18 23	18 21	18 18	18 15	18 12	18 10	18 07	18 05	18 02	N 20
N 10	32	30	28	26	24	22	21	19	17	15	13	11	10	08	06	N 10
0	25	24	23	22	21	20	19	18	16	15	14	13	13	12	11	0
S 10	19	19	18	18	18	18	17	17	17	17	17	16	16	16	16	S 10
20	13	14	14	15	16	16	17	18	18	19	20	20	21	22	23	20
S 30	18 08	18 09	18 11	18 12	18 14	18 16	18 17	18 19	18 20	18 22	18 24	18 26	18 27	18 29	18 31	S 30
35	05	07	09	11	13	16	18	20	22	24	27	29	32	34	37	35
40	18 02	05	07	10	13	16	18	21	24	27	30	33	36	39	43	40
45	17 59	18 02	06	09	13	16	20	23	27	31	34	38	42	46	50	45
50	55	17 59	04	08	12	17	21	26	31	35	40	45	49	54	18 59	50
S 52	17 54	17 58	18 03	18 08	18 12	18 17	18 22	18 27	18 32	18 37	18 42	18 47	18 53	18 58	19 04	S 52
54	52	57	02	07	13	18	23	29	34	40	45	51	18 57	19 03	09	54
56	50	56	01	07	13	18	24	30	36	42	48	55	19 01	08	14	56
58	48	54	01	07	13	19	26	32	39	45	52	18 59	06	13	20	58
S 60	17 46	17 53	18 00	18 06	18 13	18 20	18 27	18 34	18 41	18 49	18 56	19 04	19 12	19 19	19 28	S 60

CONVERSION OF ARC TO TIME

°	h m	°	h m	°	h m	°	h m	°	h m	°	h m	′	m s
0	0 00	60	4 00	120	8 00	180	12 00	240	16 00	300	20 00	0	0 00
1	0 04	61	4 04	121	8 04	181	12 04	241	16 04	301	20 04	1	0 04
2	0 08	62	4 08	122	8 08	182	12 08	242	16 08	302	20 08	2	0 08
3	0 12	63	4 12	123	8 12	183	12 12	243	16 12	303	20 12	3	0 12
4	0 16	64	4 16	124	8 16	184	12 16	244	16 16	304	20 16	4	0 16
5	0 20	65	4 20	125	8 20	185	12 20	245	16 20	305	20 20	5	0 20
6	0 24	66	4 24	126	8 24	186	12 24	246	16 24	306	20 24	6	0 24
7	0 28	67	4 28	127	8 28	187	12 28	247	16 28	307	20 28	7	0 28
8	0 32	68	4 32	128	8 32	188	12 32	248	16 32	308	20 32	8	0 32
9	0 36	69	4 36	129	8 36	189	12 36	249	16 36	309	20 36	9	0 36
10	0 40	70	4 40	130	8 40	190	12 40	250	16 40	310	20 40	10	0 40
11	0 44	71	4 44	131	8 44	191	12 44	251	16 44	311	20 44	11	0 44
12	0 48	72	4 48	132	8 48	192	12 48	252	16 48	312	20 48	12	0 48
13	0 52	73	4 52	133	8 52	193	12 52	253	16 52	313	20 52	13	0 52
14	0 56	74	4 56	134	8 56	194	12 56	254	16 56	314	20 56	14	0 56
15	1 00	75	5 00	135	9 00	195	13 00	255	17 00	315	21 00	15	1 00
16	1 04	76	5 04	136	9 04	196	13 04	256	17 04	316	21 04	16	1 04
17	1 08	77	5 08	137	9 08	197	13 08	257	17 08	317	21 08	17	1 08
18	1 12	78	5 12	138	9 12	198	13 12	258	17 12	318	21 12	18	1 12
19	1 16	79	5 16	139	9 16	199	13 16	259	17 16	319	21 16	19	1 16
20	1 20	80	5 20	140	9 20	200	13 20	260	17 20	320	21 20	20	1 20
21	1 24	81	5 24	141	9 24	201	13 24	261	17 24	321	21 24	21	1 24
22	1 28	82	5 28	142	9 28	202	13 28	262	17 28	322	21 28	22	1 28
23	1 32	83	5 32	143	9 32	203	13 32	263	17 32	323	21 32	23	1 32
24	1 36	84	5 36	144	9 36	204	13 36	264	17 36	324	21 36	24	1 36
25	1 40	85	5 40	145	9 40	205	13 40	265	17 40	325	21 40	25	1 40
26	1 44	86	5 44	146	9 44	206	13 44	266	17 44	326	21 44	26	1 44
27	1 48	87	5 48	147	9 48	207	13 48	267	17 48	327	21 48	27	1 48
28	1 52	88	5 52	148	9 52	208	13 52	268	17 52	328	21 52	28	1 52
29	1 56	89	5 56	149	9 56	209	13 56	269	17 56	329	21 56	29	1 56
30	2 00	90	6 00	150	10 00	210	14 00	270	18 00	330	22 00	30	2 00
31	2 04	91	6 04	151	10 04	211	14 04	271	18 04	331	22 04	31	2 04
32	2 08	92	6 08	152	10 08	212	14 08	272	18 08	332	22 08	32	2 08
33	2 12	93	6 12	153	10 12	213	14 12	273	18 12	333	22 12	33	2 12
34	2 16	94	6 16	154	10 16	214	14 16	274	18 16	334	22 16	34	2 16
35	2 20	95	6 20	155	10 20	215	14 20	275	18 20	335	22 20	35	2 20
36	2 24	96	6 24	156	10 24	216	14 24	276	18 24	336	22 24	36	2 24
37	2 28	97	6 28	157	10 28	217	14 28	277	18 28	337	22 28	37	2 28
38	2 32	98	6 32	158	10 32	218	14 32	278	18 32	338	22 32	38	2 32
39	2 36	99	6 36	159	10 36	219	14 36	279	18 36	339	22 36	39	2 36
40	2 40	100	6 40	160	10 40	220	14 40	280	18 40	340	22 40	40	2 40
41	2 44	101	6 44	161	10 44	221	14 44	281	18 44	341	22 44	41	2 44
42	2 48	102	6 48	162	10 48	222	14 48	282	18 48	342	22 48	42	2 48
43	2 52	103	6 52	163	10 52	223	14 52	283	18 52	343	22 52	43	2 52
44	2 56	104	6 56	164	10 56	224	14 56	284	18 56	344	22 56	44	2 56
45	3 00	105	7 00	165	11 00	225	15 00	285	19 00	345	23 00	45	3 00
46	3 04	106	7 04	166	11 04	226	15 04	286	19 04	346	23 04	46	3 04
47	3 08	107	7 08	167	11 08	227	15 08	287	19 08	347	23 08	47	3 08
48	3 12	108	7 12	168	11 12	228	15 12	288	19 12	348	23 12	48	3 12
49	3 16	109	7 16	169	11 16	229	15 16	289	19 16	349	23 16	49	3 16
50	3 20	110	7 20	170	11 20	230	15 20	290	19 20	350	23 20	50	3 20
51	3 24	111	7 24	171	11 24	231	15 24	291	19 24	351	23 24	51	3 24
52	3 28	112	7 28	172	11 28	232	15 28	292	19 28	352	23 28	52	3 28
53	3 32	113	7 32	173	11 32	233	15 32	293	19 32	353	23 32	53	3 32
54	3 36	114	7 36	174	11 36	234	15 36	294	19 36	354	23 36	54	3 36
55	3 40	115	7 40	175	11 40	235	15 40	295	19 40	355	23 40	55	3 40
56	3 44	116	7 44	176	11 44	236	15 44	296	19 44	356	23 44	56	3 44
57	3 48	117	7 48	177	11 48	237	15 48	297	19 48	357	23 48	57	3 48
58	3 52	118	7 52	178	11 52	238	15 52	298	19 52	358	23 52	58	3 52
59	3 56	119	7 56	179	11 56	239	15 56	299	19 56	359	23 56	59	3 56

The above table is for converting expressions in arc to their equivalent in time ; its main use in this Almanac is for the conversion of longitude for application to L.M.T. (*added* if *west*, *subtracted* if *east*) to give G.M.T., or vice versa, particularly in the case of sunrise, sunset, etc.

A82

POLARIS (POLE STAR) TABLE, 1966

FOR DETERMINING THE LATITUDE FROM A SEXTANT ALTITUDE

L.H.A. ♈	Q	L.H.A. ♈	Q	L.H.A. ♈	Q	L.H.A. ♈	Q	L.H.A. ♈	Q	L.H.A. ♈	Q	L.H.A. ♈	Q	L.H.A. ♈	Q
° ′	′	° ′	′	° ′	′	° ′	′	° ′	′	° ′	′	° ′	′	° ′	′
358 40	-46	80 47	-33	112 33	-6	142 10	+21	182 36	+48	264 12	+31	295 36	+4	325 19	-23
0 46	-47	82 09	-32	113 38	-5	143 20	+22	185 04	+49	265 31	+30	296 40	+3	326 30	-24
3 01	-48	83 29	-31	114 42	-4	144 31	+23	187 48	+50	266 49	+29	297 45	+2	327 42	-25
5 27	-49	84 47	-30	115 47	-3	145 42	+24	190 53	+51	268 06	+28	298 49	+1	328 55	-26
8 08	-50	86 05	-29	116 51	-2	146 55	+25	194 33	+52	269 22	+27	299 54	0	330 08	-27
11 10	-51	87 21	-28	117 56	-1	148 08	+26	199 23	+53	270 37	+26	300 59	-1	331 22	-28
14 47	-52	88 37	-27	119 00	0	149 22	+27	220 36	+52	271 51	+25	302 03	-2	332 38	-29
19 33	-53	89 51	-26	120 05	+1	150 37	+28	225 26	+51	273 04	+24	303 08	-3	333 54	-30
40 26	-52	91 04	-25	121 10	+2	151 53	+29	229 06	+50	274 17	+23	304 12	-4	335 12	-31
45 12	-51	92 17	-24	122 14	+3	153 10	+30	232 11	+49	275 28	+22	305 17	-5	336 30	-32
48 49	-50	93 29	-23	123 19	+4	154 28	+31	234 55	+48	276 39	+21	306 21	-6	337 50	-33
51 51	-49	94 40	-22	124 23	+5	155 47	+32	237 23	+47	277 49	+20	307 26	-7	339 12	-34
54 32	-48	95 50	-21	125 28	+6	157 08	+33	239 39	+46	278 59	+19	308 31	-8	340 35	-35
56 58	-47	97 00	-20	126 33	+7	158 30	+34	241 47	+45	280 08	+18	309 36	-9	342 00	-36
59 13	-46	98 09	-19	127 38	+8	159 54	+35	243 47	+44	281 17	+17	310 41	-10	343 26	-37
61 19	-45	99 18	-18	128 43	+9	161 20	+36	245 42	+43	282 25	+16	311 47	-11	344 55	-38
63 18	-44	100 26	-17	129 49	+10	162 48	+37	247 31	+42	283 32	+15	312 52	-12	346 26	-39
65 11	-43	101 34	-16	130 54	+11	164 17	+38	249 16	+41	284 39	+14	313 58	-13	348 00	-40
66 59	-42	102 41	-15	132 00	+12	165 49	+39	250 57	+40	285 46	+13	315 04	-14	349 37	-41
68 42	-41	103 48	-14	133 06	+13	167 24	+40	252 35	+39	286 53	+12	316 11	-15	351 17	-42
70 22	-40	104 55	-13	134 13	+14	169 02	+41	254 10	+38	287 59	+11	317 18	-16	353 00	-43
71 59	-39	106 01	-12	135 20	+15	170 43	+42	255 42	+37	289 05	+10	318 25	-17	354 48	-44
73 33	-38	107 07	-11	136 27	+16	172 28	+43	257 11	+36	290 10	+9	319 33	-18	356 41	-45
75 04	-37	108 12	-10	137 34	+17	174 17	+44	258 39	+35	291 16	+8	320 41	-19	358 40	-46
76 33	-36	109 18	-9	138 42	+18	176 12	+45	260 05	+34	292 21	+7	321 50	-20	0 46	-47
77 59	-35	110 23	-8	139 51	+19	178 12	+46	261 29	+33	293 26	+6	322 59	-21	3 01	-48
79 24	-34	111 28	-7	141 00	+20	180 20	+47	262 51	+32	294 31	+5	324 09	-22	5 27	-49
80 47		112 33		142 10		182 36		264 12		295 36		325 19		8 08	

Q, which does *not* include refraction, is to be applied to the corrected sextant altitude of *Polaris*.
Polaris: Mag. 2·1, S.H.A. 330°00′ ; Dec. N. 89°06′.6

STANDARD DOME REFRACTION

To be *subtracted* from sextant altitude when using sextant suspension in a perspex dome

Alt.	Refn.	Alt.	Refn.
°	′	°	′
10	8	50	4
20	7	60	4
30	6	70	3
40	5	80	3

This table must not be used if a calibration table is fitted to the dome, or if a flat glass plate is provided, or for non-standard domes.

BUBBLE SEXTANT ERROR

Sextant No.	Alt.	Corr.
	°	′

L.H.A. ♈ 300°—120°	AZIMUTH OF *POLARIS* Latitude 0°	30°	50°	55°	60°	65°	70°	L.H.A. ♈ 120°—300°
°	°	°	°	°	°	°	°	°
300	0·9	1·0	1·4	1·5	1·8	2·1	2·6	300
310	0·9	1·0	1·4	1·5	1·8	2·1	2·6	290
320	0·8	1·0	1·3	1·5	1·7	2·0	2·4	280
330	0·8	0·9	1·2	1·3	1·5	1·8	2·3	270
340	0·7	0·8	1·1	1·2	1·4	1·6	2·0	260
350	0·6	0·7	0·9	1·0	1·1	1·4	1·7	250
0	0·4	0·5	0·7	0·8	0·9	1·1	1·3	240
10	0·3	0·3	0·5	0·5	0·6	0·7	0·9	230
20	0·2	0·2	0·2	0·3	0·3	0·4	0·5	220
30	0·0	0·0	0·0	0·0	0·0	0·0	0·0	210
40	359·8	359·8	359·8	359·7	359·7	359·6	359·5	200
50	359·7	359·7	359·5	359·5	359·4	359·3	359·1	190
60	359·6	359·5	359·3	359·2	359·1	358·9	358·7	180
70	359·4	359·3	359·1	359·0	358·9	358·6	358·3	170
80	359·3	359·2	358·9	358·8	358·6	358·4	358·0	160
90	359·2	359·1	358·8	358·7	358·5	358·2	357·7	150
100	359·2	359·0	358·7	358·5	358·3	358·0	357·6	140
110	359·1	359·0	358·6	358·5	358·2	357·9	357·4	130
120	359·1	359·0	358·6	358·5	358·2	357·9	357·4	120

When Cassiopeia is left (right), *Polaris* is west (east).

CORRECTIONS TO BE APPLIED TO SEXTANT ALTITUDE

REFRACTION

To be subtracted from sextant altitude (referred to as observed altitude in A.P. 3270).

R_o	Height above sea level in units of 1,000 ft. — Sextant Altitude: 0	5	10	15	20	25	30	35	40	45	50	55	R_o	$R = R_o \times f$, f = 0·9	1·0	1·1	1·2
	° ′	° ′	° ′	° ′	° ′	° ′	° ′	° ′	° ′	° ′	° ′	° ′		R			
′	90	90	90	90	90	90	90	90	90	90	90	90	′	′	′	′	′
0													0	0	0	0	0
	63	59	55	51	46	41	36	31	26	20	17	13					
1													1	1	1	1	1
	33	29	26	22	19	16	14	11	9	7	6	4					
2													2	2	2	2	2
	21	19	16	14	12	10	8	7	5	4	2 40	1 40					
3													3	3	3	3	4
	16	14	12	10	8	7	6	5	3 10	2 20	1 30	0 40					
4													4	4	4	4	5
	12	11	9	8	7	5	4 00	3 10	2 10	1 30	0 39	+0 05					
5													5	5	5	5	6
	10	9	7	5 50	4 50	3 50	3 10	2 20	1 30	0 49	+0 11	−0 19					
6													6	5	6	7	7
	8 10	6 50	5 50	4 50	4 00	3 00	2 20	1 50	1 10	0 24	−0 11	−0 38					
7													7	6	7	8	8
	6 50	5 50	5 00	4 00	3 10	2 30	1 50	1 20	0 38	+0 04	−0 28	−0 54					
8													8	7	8	9	10
	6 00	5 10	4 10	3 20	2 40	2 00	1 30	1 00	0 19	−0 13	−0 42	−1 08					
9													9	8	9	10	11
	5 20	4 30	3 40	2 50	2 10	1 40	1 10	0 35	+0 03	−0 27	−0 53	−1 18					
10													10	9	10	11	12
	4 30	3 40	2 50	2 20	1 40	1 10	0 37	+0 11	−0 16	−0 43	−1 08	−1 31					
12													12	11	12	13	14
	3 30	2 50	2 10	1 40	1 10	0 34	+0 09	−0 14	−0 37	−1 00	−1 23	−1 44					
14													14	13	14	15	17
	2 50	2 10	1 40	1 10	0 37	+0 10	−0 13	−0 34	−0 53	−1 14	−1 35	−1 56					
16													16	14	16	18	19
	2 20	1 40	1 20	0 43	+0 15	−0 08	−0 31	−0 52	−1 08	−1 27	−1 46	−2 05					
18													18	16	18	20	22
	1 50	1 20	0 49	+0 23	−0 02	−0 26	−0 46	−1 06	−1 22	−1 39	−1 57	−2 14					
20													20	18	20	22	24
	1 12	0 44	+0 19	−0 06	−0 28	−0 48	−1 09	−1 27	−1 42	−1 58	−2 14	−2 30					
25													25	22	25	28	30
	0 34	+0 10	−0 13	−0 36	−0 55	−1 14	−1 32	−1 51	−2 06	−2 21	−2 34	−2 49					
30													30	27	30	33	36
	+0 06	−0 16	−0 37	−0 59	−1 17	−1 33	−1 51	−2 07	−2 23	−2 37	−2 51	−3 04					
35													35	31	35	38	42
	−0 18	−0 37	−0 58	−1 16	−1 34	−1 49	−2 06	−2 22	−2 35	−2 49	−3 03	−3 16					
40													40	36	40	44	48
		−0 53	−1 14	−1 31	−1 47	−2 03	−2 18	−2 33	−2 47	−2 59	−3 13	−3 25					
45													45	40	45	50	54
		−1 10	−1 28	−1 44	−1 59	−2 15	−2 28	−2 43	−2 56	−3 08	−3 22	−3 33					
50													50	45	50	55	60
			−1 40	−1 53	−2 09	−2 24	−2 38	−2 52	−3 04	−3 17	−3 29	−3 41					
55													55	49	55	60	66
				−2 03	−2 18	−2 33	−2 46	−3 01	−3 12	−3 25	−3 37	−3 48					
60													60	54	60	66	72
							−2 53	−3 07	−3 19	−3 31	−3 42	−3 53					

f	Height — Temperature in °C.: 0	5	10	15	20	25	30	35	40	45	50	55	f	f = 0·9	1·0	1·1	1·2
	+47	+36	+27	+18	+10	+ 3	− 5	−13	For these heights no temperature correction is necessary, so use $R=R_o$					When R_o is less than 10′ or the height greater than 35,000 ft. use $R=R_o$			
0·9													0·9				
	+26	+16	+ 6	− 4	−13	−22	−31	−40									
1·0													1·0				
	+ 5	− 5	−15	−25	−36	−46	−57	−68									
1·1													1·1				
	−16	−25	−36	−46	−58	−71	−83	−95									
1·2													1·2				
	−37	−45	−56	−67	−81	−95											

Choose the column appropriate to height, in units of 1,000 ft., and find the range of altitude in which the sextant altitude lies ; the corresponding value of R_o is the refraction, to be subtracted from sextant altitude, unless conditions are extreme. In that case find f from the lower table, with critical argument temperature. Use the table on the right to form the refraction, $R=R_o \times f$.

CORIOLIS (Z) CORRECTION

To be applied by moving the position line a distance Z to starboard (right) of the track in northern latitudes and to port (left) in southern latitudes. The argument is given as T.A.S. (True Air Speed) in A.P. 3270.

G/S KNOTS	Latitude 0°	10°	20°	30°	40°	50°	60°	70°	80°	90°	G/S KNOTS	Latitude 0°	10°	20°	30°	40°	50°	60°	70°	80°	90°
	′	′	′	′	′	′	′	′	′	′		′	′	′	′	′	′	′	′	′	′
150	0	1	1	2	3	3	3	4	4	4	550	0	3	5	7	9	11	12	14	14	14
200	0	1	2	3	3	4	5	5	5	5	600	0	3	5	8	10	12	14	15	16	16
250	0	1	2	3	4	5	6	6	6	7	650	0	3	6	9	11	13	15	16	17	17
300	0	1	3	4	5	6	7	7	8	8	700	0	3	6	9	12	14	16	17	18	18
350	0	2	3	5	6	7	8	9	9	9	750	0	3	7	10	13	15	17	18	19	20
400	0	2	4	5	7	8	9	10	10	10	800	0	4	7	10	13	16	18	20	21	21
450	0	2	4	6	8	9	10	11	12	12	850	0	4	8	11	14	17	19	21	22	22
500	0	2	4	7	8	10	11	12	13	13	900	0	4	8	12	15	18	20	22	23	24

APP. B — AIR ALMANAC

CORRECTIONS TO BE APPLIED TO MARINE SEXTANT ALTITUDES

MARINE SEXTANT ERROR

Sextant No.

Index Error

CORRECTIONS

In addition to sextant error and dip, corrections are to be applied for:

- Refraction
- Semi-diameter (for the Sun and Moon)
- Parallax (for the Moon)
- Dome refraction if applicable.

CORRECTION FOR DIP OF THE HORIZON

To be subtracted from sextant altitude

Ht.	Dip	Ht.	Dip	Ht.	Dip	Ht.	Dip	Ht.	Dip
Ft.	′	Ft.	′	Ft.	′	Ft.	′	Ft.	′
0		114		437		968		1,707	
	1		11		21		31		41
2		137		481		1,033		1,792	
	2		12		22		32		42
6		162		527		1,099		1,880	
	3		13		23		33		43
12		189		575		1,168		1,970	
	4		14		24		34		44
21		218		625		1,239		2,061	
	5		15		25		35		45
31		250		677		1,311		2,155	
	6		16		26		36		46
43		283		731		1,386		2,251	
	7		17		27		37		47
58		318		787		1,463		2,349	
	8		18		28		38		48
75		356		845		1,543		2,449	
	9		19		29		39		49
93		395		906		1,624		2,551	
	10		20		30		40		50
114		437		968		1,707		2,655	

LIST OF CONTENTS

Pages	Contents
Inside front cover	Star list (57 stars) and G.H.A. interpolation tables
Daily pages	Ephemerides of Sun, Moon, Aries and planets ; moonrise and moonset
F1—F2 (flap)	Star chart
F3 (flap)	Star list (57 stars) and G.H.A. interpolation tables
F4 (flap)	Interpolation of moonrise and moonset for longitude, and star index
A1—A3	Title page, preface, etc.
A4—A17	Explanation
A18—A19	List of symbols and abbreviations
A20—A23	Standard times
A24—A57	Sky diagrams
A58—A59	Planet location diagram
A60—A61	Astrograph settings
A62—A67	Sunrise, sunset and civil twilight
A68—A73	Rising, setting and depression graphs
A74—A75	Semi-duration graphs of sunlight, twilight and moonlight, in high latitudes
A76—A78	Star list, 173 stars, (accuracy 0′·1)
A79	Conversion of arc to time
A80—A81	Interpolation of G.H.A. Sun and G.H.A. Aries (accuracy 0′·1)
A82	*Polaris* tables and dome refraction
Inside back cover	Corrections for (total) refraction and Coriolis (Z) table
Outside back cover	Corrections to marine sextant observations

APPENDIX C — H.O. 249, VOL. I

LAT 30°N

LHA ♈	Hc	Zn	Hc	Zn	Hc	Zn	Hc	Zn	Hc	Zn	Hc	Zn	Hc	Zn
	* VEGA		Alphecca		* SPICA		Gienah		REGULUS		* POLLUX		Dubhe	
180	12 00	052	43 26	080	44 22	151	42 32	176	58 13	242	34 37	285	56 40	348
181	12 41	052	44 17	081	44 47	152	42 35	177	57 27	243	33 46	286	56 28	347
182	13 23	053	45 08	081	45 11	153	42 38	178	56 41	244	32 56	286	56 16	346
183	14 04	053	46 00	081	45 34	155	42 39	179	55 54	245	32 07	286	56 03	345
184	14 46	053	46 51	082	45 56	156	42 39	181	55 06	246	31 17	287	55 49	345
185	15 28	054	47 43	082	46 17	157	42 38	182	54 19	247	30 27	287	55 35	344
186	16 10	054	48 34	082	46 36	159	42 35	183	53 31	248	29 37	287	55 20	343
187	16 52	055	49 26	083	46 55	160	42 32	185	52 42	249	28 48	288	55 05	342
188	17 34	055	50 17	083	47 12	161	42 27	186	51 54	250	27 58	288	54 49	342
189	18 17	055	51 09	083	47 28	163	42 21	187	51 05	251	27 09	288	54 32	341
190	19 00	056	52 00	084	47 43	164	42 14	188	50 15	252	26 20	289	54 15	341
191	19 43	056	52 52	084	47 56	166	42 06	190	49 26	252	25 31	289	53 58	340
192	20 26	056	53 44	085	48 09	167	41 57	191	48 36	253	24 41	290	53 39	339
193	21 09	057	54 36	085	48 20	168	41 47	192	47 46	254	23 53	290	53 21	339
194	21 52	057	55 27	085	48 30	170	41 35	193	46 56	255	23 04	290	53 01	338
	* Kochab		VEGA		Rasalhague		* ANTARES		SPICA		* REGULUS		Dubhe	
195	43 32	010	22 36	057	24 54	089	14 56	133	48 38	171	46 06	256	52 42	338
196	43 41	010	23 20	057	25 46	090	15 34	134	48 45	173	45 16	256	52 22	337
197	43 50	009	24 04	058	26 38	091	16 11	135	48 51	174	44 25	257	52 01	336
198	43 58	009	24 48	058	27 30	091	16 48	135	48 56	176	43 35	258	51 40	336
199	44 06	009	25 32	058	28 22	092	17 24	136	48 59	177	42 44	258	51 18	335
200	44 14	008	26 16	059	29 14	092	18 00	137	49 01	179	41 53	259	50 57	335
201	44 21	008	27 00	059	30 06	093	18 36	137	49 01	180	41 02	260	50 34	335
202	44 28	008	27 45	059	30 58	093	19 11	138	49 00	182	40 11	260	50 12	334
203	44 35	007	28 30	059	31 50	094	19 45	139	48 58	183	39 19	261	49 49	334
204	44 42	007	29 14	060	32 41	094	20 19	140	48 54	185	38 28	261	49 26	333
205	44 48	007	29 59	060	33 33	095	20 53	140	48 50	186	37 37	262	49 02	333
206	44 54	006	30 44	060	34 25	095	21 26	141	48 43	188	36 45	263	48 38	333
207	44 59	006	31 29	060	35 17	096	21 58	142	48 36	189	35 53	263	48 14	332
208	45 05	006	32 15	061	36 08	096	22 30	143	48 27	191	35 02	264	47 49	332
209	45 09	005	33 00	061	37 00	097	23 01	143	48 16	192	34 10	264	47 25	331
	Kochab		* VEGA		Rasalhague		ANTARES		* SPICA		REGULUS		* Dubhe	
210	45 14	005	33 45	061	37 52	098	23 32	144	48 05	194	33 18	265	47 00	331
211	45 18	004	34 31	061	38 43	098	24 02	145	47 52	195	32 27	265	46 34	331
212	45 22	004	35 17	062	39 35	099	24 32	146	47 38	196	31 35	266	46 09	331
213	45 26	004	36 02	062	40 26	099	25 01	147	47 23	198	30 43	267	45 43	330
214	45 29	003	36 48	062	41 17	100	25 29	148	47 06	199	29 51	267	45 17	330
215	45 32	003	37 34	062	42 08	101	25 56	148	46 49	201	28 59	268	44 51	330
216	45 34	003	38 20	062	42 59	101	26 23	149	46 30	202	28 07	268	44 25	330
217	45 36	002	39 06	062	43 50	102	26 50	150	46 10	203	27 15	269	43 58	329
218	45 38	002	39 52	063	44 41	103	27 15	151	45 49	205	26 23	269	43 31	329
219	45 40	001	40 38	063	45 32	103	27 40	152	45 27	206	25 31	270	43 04	329
220	45 41	001	41 25	063	46 22	104	28 04	153	45 04	207	24 39	270	42 37	329
221	45 42	001	42 11	063	47 12	105	28 28	154	44 39	208	23 48	271	42 10	329
222	45 42	000	42 57	063	48 03	106	28 50	155	44 14	210	22 56	271	41 43	328
223	45 42	000	43 44	063	48 53	106	29 12	156	43 48	211	22 04	272	41 16	328
224	45 42	000	44 30	064	49 42	107	29 34	157	43 21	212	21 12	272	40 48	328
	VEGA		* ALTAIR		Nunki		ANTARES		* SPICA		Denebola		* Alkaid	
225	45 17	064	19 41	091	10 44	129	29 54	158	42 53	213	43 18	262	66 02	330
226	46 03	064	20 33	092	11 24	130	30 13	158	42 24	214	42 27	263	65 35	329
227	46 50	064	21 25	092	12 04	130	30 32	159	41 54	216	41 35	263	65 07	327
228	47 37	064	22 17	093	12 43	131	30 50	160	41 23	217	40 43	264	64 39	326
229	48 23	064	23 09	093	13 22	132	31 07	161	40 52	218	39 52	264	64 10	325
230	49 10	064	24 01	094	14 01	132	31 23	162	40 20	219	39 00	265	63 40	324
231	49 57	064	24 53	094	14 39	133	31 39	163	39 47	220	38 08	265	63 09	324
232	50 44	064	25 45	095	15 17	134	31 53	164	39 13	221	37 16	266	62 38	323
233	51 30	064	26 36	095	15 55	134	32 07	165	38 39	222	36 24	267	62 06	322
234	52 17	064	27 28	096	16 32	135	32 19	166	38 04	223	35 33	267	61 34	321
235	53 04	064	28 20	096	17 08	136	32 31	168	37 28	224	34 41	268	61 01	320
236	53 51	064	29 11	097	17 44	136	32 42	169	36 51	225	33 49	268	60 27	320
237	54 37	064	30 03	098	18 20	137	32 52	170	36 14	226	32 57	269	59 53	319
238	55 24	064	30 54	098	18 55	138	33 01	171	35 37	227	32 05	269	59 19	318
239	56 11	064	31 46	099	19 30	139	33 09	172	34 59	228	31 13	270	58 44	318
	* VEGA		ALTAIR		Nunki		* ANTARES		SPICA		ARCTURUS		* Alkaid	
240	56 58	064	32 37	099	20 04	139	33 16	173	34 20	229	63 47	252	58 09	317
241	57 44	064	33 28	100	20 38	140	33 22	174	33 40	230	62 57	253	57 34	317
242	58 31	064	34 20	101	21 11	141	33 27	175	33 01	230	62 07	254	56 58	316
243	59 18	064	35 11	101	21 44	142	33 32	176	32 20	231	61 17	255	56 22	316
244	60 04	064	36 02	102	22 16	142	33 35	177	31 40	232	60 27	256	55 46	315
245	60 51	063	36 52	102	22 48	143	33 37	178	30 58	233	59 36	257	55 09	315
246	61 37	063	37 43	103	23 19	144	33 38	179	30 17	234	58 45	258	54 32	315
247	62 23	063	38 34	104	23 49	145	33 39	180	29 35	235	57 55	258	53 55	314
248	63 10	063	39 24	104	24 19	145	33 38	181	28 52	235	57 04	259	53 18	314
249	63 56	062	40 14	105	24 48	146	33 36	182	28 09	236	56 13	260	52 40	314
250	64 42	062	41 04	106	25 17	147	33 34	183	27 26	237	55 21	261	52 03	314
251	65 27	062	41 54	106	25 45	148	33 30	184	26 42	238	54 30	261	51 25	313
252	66 13	061	42 44	107	26 12	149	33 26	186	25 58	238	53 39	262	50 47	313
253	66 58	061	43 34	108	26 38	150	33 20	187	25 14	239	52 47	263	50 09	313
254	67 44	060	44 23	109	27 04	151	33 14	188	24 29	240	51 56	263	49 31	313
	* DENEB		ALTAIR		Nunki		* ANTARES		SPICA		ARCTURUS		* Alkaid	
255	44 46	055	45 12	109	27 30	152	33 06	189	23 44	240	51 04	264	48 52	312
256	45 29	055	46 01	110	27 54	152	32 58	190	22 59	241	50 12	265	48 14	312
257	46 11	054	46 50	111	28 18	153	32 48	191	22 13	242	49 20	265	47 35	312
258	46 53	054	47 38	112	28 41	154	32 38	192	21 27	242	48 29	266	46 57	312
259	47 35	054	48 26	113	29 03	155	32 27	193	20 41	243	47 37	266	46 18	312
260	48 18	054	49 14	114	29 25	156	32 15	194	19 55	244	46 45	267	45 39	312
261	49 00	054	50 02	115	29 45	157	32 02	195	19 08	244	45 53	267	45 00	312
262	49 42	054	50 49	116	30 05	158	31 48	196	18 21	245	45 01	268	44 21	312
263	50 24	054	51 35	117	30 24	159	31 33	197	17 34	246	44 09	268	43 42	312
264	51 06	054	52 22	118	30 43	160	31 17	198	16 46	246	43 17	269	43 04	312
265	51 48	054	53 08	119	31 00	161	31 01	199	15 59	247	42 25	269	42 25	312
266	52 30	054	53 53	120	31 17	162	30 44	200	15 11	247	41 33	270	41 46	312
267	53 12	053	54 38	121	31 33	163	30 25	201	14 23	248	40 41	271	41 07	312
268	53 54	053	55 23	122	31 47	164	30 06	202	13 35	249	39 49	271	40 28	312
269	54 35	053	56 06	123	32 01	165	29 46	203	12 46	249	38 57	272	39 49	312

LAT 30°N

LHA ♈	Hc	Zn	Hc	Zn	Hc	Zn	Hc	Zn	Hc	Zn	Hc	Zn	Hc	Zn
	* DENEB		ALTAIR		* Nunki		ANTARES		* ARCTURUS		Alkaid		Kochab	
270	55 17	053	56 50	124	32 15	166	29 26	204	38 06	272	39 10	312	39 49	345
271	55 58	053	57 32	126	32 27	167	29 04	205	37 14	273	38 31	312	39 35	345
272	56 39	052	58 14	127	32 38	168	28 42	206	36 22	273	37 52	312	39 21	345
273	57 20	052	58 56	128	32 48	169	28 19	207	35 30	273	37 13	312	39 08	345
274	58 01	052	59 36	130	32 58	170	27 55	208	34 38	274	36 34	312	38 54	344
275	58 42	051	60 16	131	33 06	171	27 31	209	33 46	274	35 55	312	38 39	344
276	59 22	051	60 55	133	33 14	172	27 06	209	32 54	275	35 17	312	38 25	344
277	60 02	050	61 33	134	33 20	173	26 40	210	32 03	275	34 38	312	38 11	344
278	60 42	050	62 09	136	33 26	174	26 13	211	31 11	276	33 59	312	37 56	344
279	61 22	049	62 45	138	33 31	176	25 46	212	30 19	276	33 21	312	37 41	344
280	62 01	049	63 20	139	33 34	177	25 18	213	29 27	277	32 42	312	37 26	343
281	62 40	048	63 53	141	33 37	178	24 50	214	28 36	277	32 04	312	37 12	343
282	63 19	048	64 25	143	33 39	179	24 21	215	27 44	278	31 25	313	36 56	343
283	63 57	047	64 56	145	33 39	180	23 51	215	26 53	278	30 47	313	36 41	343
284	64 34	046	65 25	147	33 39	181	23 21	216	26 01	278	30 09	313	36 26	343
	* Alpheratz		Enif		ALTAIR		* ANTARES		ARCTURUS		* Alkaid		Kochab	
285	24 38	070	47 05	110	65 52	149	22 50	217	25 10	279	29 31	313	36 11	343
286	25 27	070	47 54	110	66 18	151	22 18	218	24 19	279	28 53	313	35 55	343
287	26 15	070	48 42	111	66 42	154	21 46	219	23 27	280	28 15	313	35 40	343
288	27 04	071	49 31	112	67 04	156	21 13	219	22 36	280	27 38	314	35 24	343
289	27 54	071	50 19	113	67 24	158	20 40	220	21 45	281	27 00	314	35 08	342
290	28 43	071	51 06	114	67 43	161	20 07	221	20 54	281	26 22	314	34 52	342
291	29 32	072	51 54	115	67 59	163	19 32	222	20 03	281	25 45	314	34 37	342
292	30 21	072	52 41	116	68 13	166	18 58	222	19 12	282	25 08	314	34 21	342
293	31 11	072	53 27	117	68 24	169	18 23	223	18 21	282	24 31	315	34 05	342
294	32 00	073	54 13	118	68 33	171	17 47	224	17 31	283	23 54	315	33 49	342
295	32 50	073	54 59	119	68 40	174	17 11	224	16 40	283	23 17	315	33 33	342
296	33 40	073	55 44	120	68 44	177	16 34	225	15 49	284	22 40	315	33 17	342
297	34 30	074	56 29	121	68 46	179	15 57	226	14 59	284	22 04	316	33 00	342
298	35 20	074	57 13	123	68 46	182	15 20	226	14 08	285	21 27	316	32 44	342
299	36 10	074	57 57	124	68 43	185	14 42	227	13 18	285	20 51	316	32 28	342
	* Schedar		Alpheratz		* FOMALHAUT		ALTAIR		Rasalhague		* Alphecca		Kochab	
300	35 41	040	37 00	075	17 01	141	68 37	187	51 54	251	32 06	285	32 12	342
301	36 14	040	37 50	075	17 33	142	68 29	190	51 04	252	31 16	285	31 56	342
302	36 47	040	38 40	075	18 05	143	68 19	193	50 15	252	30 26	285	31 39	342
303	37 21	040	39 30	076	18 37	143	68 06	195	49 25	253	29 35	286	31 23	342
304	37 54	040	40 21	076	19 08	144	67 52	198	48 35	254	28 46	286	31 07	342
305	38 27	040	41 11	076	19 38	145	67 35	200	47 45	255	27 56	287	30 51	342
306	39 00	040	42 02	077	20 08	145	67 15	203	46 55	256	27 06	287	30 35	342
307	39 33	040	42 52	077	20 37	146	66 54	205	46 05	256	26 16	287	30 18	342
308	40 07	040	43 43	077	21 06	147	66 31	208	45 14	257	25 27	288	30 02	342
309	40 40	040	44 34	078	21 34	148	66 06	210	44 23	258	24 37	288	29 46	342
310	41 13	039	45 24	078	22 01	149	65 40	212	43 33	258	23 48	288	29 30	342
311	41 46	039	46 15	078	22 28	149	65 11	214	42 42	259	22 58	289	29 14	342
312	42 19	039	47 06	079	22 54	150	64 42	216	41 51	260	22 09	289	28 58	342
313	42 52	039	47 57	079	23 20	151	64 10	218	40 59	260	21 20	290	28 42	342
314	43 24	039	48 48	079	23 45	152	63 38	220	40 08	261	20 31	290	28 26	342
	* Mirfak		Hamal		Diphda		* FOMALHAUT		ALTAIR		* VEGA		Kochab	
315	19 10	043	22 43	075	18 06	125	24 09	153	63 04	222	59 12	296	28 10	342
316	19 46	043	23 33	076	18 48	125	24 33	154	62 29	223	58 25	296	27 54	342
317	20 21	043	24 24	076	19 30	126	24 55	154	61 52	225	57 39	296	27 38	342
318	20 57	044	25 14	077	20 12	127	25 18	155	61 15	227	56 52	296	27 22	342
319	21 33	044	26 05	077	20 54	127	25 39	156	60 37	228	56 05	296	27 06	343
320	22 09	044	26 55	077	21 35	128	26 00	157	59 58	230	55 18	296	26 51	343
321	22 46	044	27 46	078	22 16	129	26 20	158	59 18	231	54 32	296	26 35	343
322	23 22	045	28 37	078	22 56	130	26 39	159	58 37	233	53 45	296	26 19	343
323	23 59	045	29 28	079	23 36	130	26 57	160	57 55	234	52 58	296	26 04	343
324	24 36	045	30 19	079	24 15	131	27 15	161	57 13	235	52 11	296	25 49	343
325	25 13	045	31 10	079	24 54	132	27 32	162	56 30	236	51 24	296	25 33	343
326	25 50	046	32 01	080	25 33	133	27 48	163	55 46	238	50 38	296	25 18	343
327	26 27	046	32 52	080	26 11	133	28 03	163	55 02	239	49 51	296	25 03	343
328	27 04	046	33 43	081	26 49	134	28 18	164	54 17	240	49 04	296	24 48	343
329	27 42	046	34 35	081	27 26	135	28 31	165	53 32	241	48 17	296	24 33	344
	* Mirfak		Hamal		Diphda		* FOMALHAUT		ALTAIR		* VEGA		Kochab	
330	28 19	046	35 26	081	28 02	136	28 44	166	52 46	242	47 31	296	24 18	344
331	28 57	047	36 17	082	28 38	137	28 56	167	52 00	243	46 44	296	24 04	344
332	29 35	047	37 09	082	29 14	137	29 07	168	51 14	244	45 58	296	23 49	344
333	30 12	047	38 00	083	29 49	138	29 18	169	50 27	245	45 11	296	23 35	344
334	30 50	047	38 52	083	30 23	139	29 27	170	49 40	246	44 24	297	23 20	344
335	31 29	047	39 43	083	30 57	140	29 35	171	48 52	247	43 38	297	23 06	344
336	32 07	047	40 35	084	31 30	141	29 43	172	48 04	248	42 52	297	22 52	344
337	32 45	047	41 27	084	32 02	142	29 50	173	47 16	249	42 05	297	22 38	345
338	33 23	048	42 19	085	32 34	143	29 56	174	46 27	249	41 19	297	22 24	345
339	34 02	048	43 10	085	33 05	144	30 00	175	45 39	250	40 33	297	22 11	345
340	34 40	048	44 02	086	33 35	145	30 04	176	44 50	251	39 46	297	21 57	345
341	35 19	048	44 54	086	34 05	146	30 08	177	44 00	252	39 00	298	21 44	345
342	35 57	048	45 46	086	34 34	147	30 10	178	43 11	253	38 14	298	21 30	345
343	36 36	048	46 38	087	35 02	148	30 11	179	42 21	253	37 28	298	21 17	346
344	37 14	048	47 30	087	35 30	149	30 11	180	41 31	254	36 42	298	21 04	346
	* CAPELLA		ALDEBARAN		Diphda		* FOMALHAUT		ALTAIR		* VEGA		Kochab	
345	18 49	047	13 39	079	35 56	150	30 11	181	40 41	255	35 57	298	20 52	346
346	19 27	047	14 30	079	36 22	151	30 10	182	39 51	255	35 11	299	20 39	346
347	20 05	048	15 21	080	36 47	152	30 07	183	39 01	256	34 25	299	20 27	346
348	20 44	048	16 12	080	37 11	153	30 04	184	38 10	257	33 40	299	20 14	346
349	21 22	048	17 03	081	37 34	154	30 00	185	37 20	257	32 54	299	20 02	347
350	22 01	049	17 54	081	37 57	155	29 55	186	36 29	258	32 09	299	19 50	347
351	22 40	049	18 46	081	38 18	156	29 49	187	35 38	259	31 24	300	19 38	347
352	23 19	049	19 37	082	38 39	157	29 42	188	34 47	259	30 39	300	19 27	347
353	23 59	049	20 28	082	38 58	159	29 34	189	33 56	260	29 54	300	19 15	347
354	24 38	050	21 20	083	39 17	160	29 26	190	33 05	260	29 09	300	19 04	348
355	25 18	050	22 12	083	39 34	161	29 16	191	32 13	261	28 24	301	18 53	348
356	25 58	050	23 03	084	39 51	162	29 06	192	31 22	262	27 39	301	18 42	348
357	26 38	050	23 55	084	40 06	163	28 55	193	30 31	262	26 55	301	18 31	348
358	27 18	050	24 47	085	40 21	164	28 42	194	29 39	263	26 11	302	18 21	349
359	27 58	051	25 38	085	40 34	166	28 30	195	28 48	263	25 26	302	18 10	349

LAT 49°N

LHA ♈	Hc	Zn	Hc	Zn	Hc	Zn	Hc	Zn	Hc	Zn	Hc	Zn	Hc	Zn
	* CAPELLA		ALDEBARAN		Hamal		* Diphda		ALTAIR		* VEGA		Kochab	
0	39 18	062	26 23	095	54 26	125	22 11	169	24 21	255	33 30	293	36 35	347
1	39 52	062	27 02	096	54 58	126	22 18	170	23 43	255	32 53	293	36 26	347
2	40 27	063	27 42	097	55 29	128	22 24	171	23 04	256	32 17	294	36 18	347
3	41 02	063	28 21	098	56 00	129	22 30	172	22 26	257	31 41	294	36 09	348
4	41 37	064	29 00	098	56 30	130	22 35	173	21 48	258	31 06	295	36 01	348
5	42 13	064	29 39	099	57 00	132	22 39	174	21 09	259	30 30	296	35 53	348
6	42 48	065	30 17	100	57 29	133	22 42	176	20 30	260	29 55	296	35 45	349
7	43 24	065	30 56	101	57 58	135	22 45	177	19 52	260	29 19	297	35 37	349
8	44 00	066	31 35	102	58 25	136	22 47	178	19 13	261	28 44	297	35 29	349
9	44 36	066	32 13	103	58 52	138	22 49	179	18 34	262	28 09	298	35 22	349
10	45 12	067	32 52	103	59 18	139	22 49	180	17 55	263	27 34	298	35 15	350
11	45 48	067	33 30	104	59 44	141	22 49	181	17 15	263	27 00	299	35 08	350
12	46 24	068	34 08	105	60 08	143	22 48	182	16 37	264	26 26	300	35 01	350
13	47 01	068	34 46	106	60 32	144	22 47	183	15 57	265	25 51	300	34 55	351
14	47 37	069	35 24	107	60 54	146	22 45	184	15 18	266	25 18	301	34 48	351
	* CAPELLA		ALDEBARAN		Hamal		* Diphda		Alpheratz		* DENEB		Kochab	
15	48 14	069	36 01	108	61 16	148	22 42	185	67 27	212	46 58	291	34 42	351
16	48 51	070	36 39	109	61 36	149	22 38	186	67 06	214	46 22	291	34 36	352
17	49 28	070	37 16	110	61 56	151	22 34	187	66 43	216	45 45	292	34 31	352
18	50 05	071	37 53	110	62 14	153	22 29	188	66 19	218	45 08	292	34 25	352
19	50 42	071	38 30	111	62 32	155	22 24	189	65 55	220	44 32	293	34 20	353
20	51 19	072	39 06	112	62 48	157	22 17	190	65 29	222	43 56	293	34 15	353
21	51 57	072	39 43	113	63 03	159	22 10	191	65 02	223	43 20	294	34 10	353
22	52 34	073	40 19	114	63 16	161	22 02	192	64 35	225	42 44	294	34 05	353
23	53 12	073	40 54	115	63 29	163	21 54	193	64 07	227	42 08	295	34 01	354
24	53 50	074	41 30	116	63 40	165	21 45	194	63 37	229	41 32	295	33 57	354
25	54 27	074	42 05	117	63 49	167	21 35	195	63 08	230	40 57	296	33 53	354
26	55 05	075	42 40	118	63 58	169	21 25	196	62 37	232	40 21	296	33 49	355
27	55 43	075	43 14	119	64 04	171	21 14	197	62 06	233	39 46	297	33 45	355
28	56 21	076	43 48	120	64 10	173	21 02	198	61 34	235	39 11	297	33 42	355
29	56 59	076	44 22	122	64 14	175	20 49	199	61 02	236	38 36	298	33 39	356
	CAPELLA		* BETELGEUSE		RIGEL		Hamal		* Alpheratz		DENEB		* Kochab	
30	57 38	077	26 02	110	18 56	129	64 17	177	60 29	238	38 01	298	33 36	356
31	58 16	077	26 39	111	19 27	130	64 18	179	59 55	239	37 26	299	33 33	356
32	58 54	078	27 16	112	19 57	131	64 18	181	59 21	240	36 52	299	33 31	357
33	59 33	078	27 52	113	20 26	132	64 16	184	58 47	242	36 18	300	33 28	357
34	60 11	079	28 28	114	20 56	132	64 13	186	58 12	243	35 44	300	33 26	357
35	60 50	079	29 04	115	21 25	133	64 08	188	57 37	244	35 10	301	33 24	358
36	61 29	080	29 40	116	21 53	134	64 02	190	57 01	245	34 36	301	33 23	358
37	62 08	080	30 15	116	22 21	135	63 54	192	56 25	246	34 02	302	33 22	358
38	62 46	081	30 50	117	22 48	136	63 46	194	55 49	247	33 29	302	33 20	359
39	63 25	081	31 25	118	23 16	137	63 35	196	55 13	249	32 56	303	33 19	359
40	64 04	082	32 00	119	23 42	138	63 24	198	54 36	250	32 23	303	33 19	359
41	64 43	083	32 34	120	24 08	139	63 11	200	53 59	251	31 50	304	33 18	000
42	65 22	083	33 08	121	24 34	140	62 57	202	53 22	252	31 18	305	33 18	000
43	66 01	084	33 41	122	24 59	141	62 41	204	52 44	253	30 45	305	33 18	000
44	66 41	084	34 15	123	25 23	142	62 25	206	52 06	254	30 13	306	33 18	000
	* Dubhe		POLLUX		BETELGEUSE		* RIGEL		Hamal		* Alpheratz		DENEB	
45	30 39	028	33 05	084	34 47	124	25 47	143	62 07	208	51 28	255	29 41	306
46	30 58	029	33 45	085	35 20	125	26 11	144	61 48	210	50 50	256	29 09	307
47	31 17	029	34 24	085	35 52	126	26 34	145	61 28	211	50 12	257	28 38	307
48	31 36	029	35 03	086	36 23	127	26 56	146	61 07	213	49 34	258	28 07	308
49	31 55	030	35 42	087	36 55	128	27 18	147	60 45	215	48 55	259	27 36	308
50	32 15	030	36 22	087	37 25	129	27 39	148	60 22	217	48 17	259	27 05	309
51	32 35	031	37 01	088	37 55	130	27 59	149	59 58	218	47 38	260	26 34	309
52	32 55	031	37 40	089	38 25	132	28 19	150	59 34	220	46 59	261	26 04	310
53	33 16	031	38 20	090	38 54	133	28 38	151	59 08	221	46 20	262	25 34	310
54	33 36	032	38 59	091	39 23	134	28 57	152	58 41	223	45 41	263	25 04	311
55	33 57	032	39 38	091	39 51	135	29 15	154	58 14	225	45 02	264	24 34	312
56	34 18	032	40 18	092	40 19	136	29 32	155	57 46	226	44 23	265	24 05	312
57	34 39	033	40 57	093	40 46	137	29 49	156	57 18	227	43 44	265	23 36	313
58	35 01	033	41 36	094	41 13	138	30 04	157	56 48	229	43 04	266	23 07	313
59	35 22	034	42 16	094	41 39	140	30 20	158	56 18	230	42 25	267	22 38	314
	* Dubhe		POLLUX		SIRIUS		* RIGEL		Hamal		* Alpheratz		DENEB	
60	35 44	034	42 55	095	14 59	140	30 34	159	55 48	232	41 46	268	22 10	314
61	36 06	034	43 34	096	15 25	140	30 48	160	55 17	233	41 06	268	21 42	315
62	36 29	035	44 13	097	15 49	141	31 01	161	54 45	234	40 27	269	21 14	316
63	36 51	035	44 52	098	16 14	142	31 13	162	54 13	236	39 48	270	20 47	316
64	37 14	035	45 31	099	16 38	143	31 25	164	53 40	237	39 08	271	20 20	317
65	37 37	036	46 10	099	17 01	144	31 36	165	53 07	238	38 29	272	19 53	317
66	38 00	036	46 49	100	17 24	145	31 46	166	52 33	239	37 50	272	19 26	318
67	38 23	036	47 28	101	17 46	146	31 55	167	51 59	240	37 10	273	19 00	318
68	38 47	037	48 06	102	18 08	147	32 04	168	51 24	242	36 31	274	18 34	319
69	39 10	037	48 45	103	18 29	148	32 11	169	50 50	243	35 52	275	18 08	320
70	39 34	037	49 23	104	18 50	149	32 18	170	50 14	244	35 12	275	17 43	320
71	39 58	038	50 01	105	19 10	150	32 25	172	49 39	245	34 33	276	17 17	321
72	40 22	038	50 39	106	19 30	151	32 30	173	49 03	246	33 54	277	16 53	321
73	40 47	038	51 17	107	19 49	152	32 35	174	48 27	247	33 15	277	16 28	322
74	41 11	039	51 55	108	20 07	153	32 38	175	47 51	248	32 36	278	16 04	323
	* Dubhe		REGULUS		PROCYON		* SIRIUS		RIGEL		* Hamal		DENEB	
75	41 36	039	17 53	092	35 06	130	20 25	154	32 41	176	47 14	249	15 40	323
76	42 01	039	18 32	093	35 36	131	20 43	155	32 44	177	46 37	250	15 17	324
77	42 26	040	19 12	094	36 06	132	20 59	156	32 45	179	46 00	251	14 54	324
78	42 51	040	19 51	094	36 35	133	21 15	157	32 46	180	45 23	252	14 31	325
79	43 16	040	20 30	095	37 03	134	21 31	158	32 45	181	44 45	253	14 08	326
80	43 41	040	21 09	096	37 32	135	21 45	159	32 44	182	44 07	254	13 46	326
81	44 07	041	21 48	097	37 59	136	22 00	160	32 42	183	43 29	255	13 25	327
82	44 33	041	22 27	097	38 26	137	22 13	161	32 40	184	42 51	256	13 03	328
83	44 59	041	23 06	098	38 53	138	22 26	162	32 36	186	42 13	257	12 42	328
84	45 25	041	23 45	099	39 19	140	22 38	163	32 32	187	41 35	258	12 22	329
85	45 51	042	24 24	100	39 44	141	22 50	164	32 27	188	40 56	258	12 01	329
86	46 17	042	25 03	101	40 09	142	23 01	165	32 21	189	40 18	259	11 41	330
87	46 44	042	25 42	102	40 33	143	23 11	166	32 14	190	39 39	260	11 22	331
88	47 10	042	26 20	102	40 56	144	23 20	167	32 07	191	39 00	261	11 03	331
89	47 37	043	26 58	103	41 19	145	23 29	168	31 59	193	38 21	262	10 44	332

LAT 49°N

LHA ♈	Hc	Zn	Hc	Zn	Hc	Zn	Hc	Zn	Hc	Zn	Hc	Zn	Hc	Zn
	* Dubhe		REGULUS		PROCYON		* SIRIUS		RIGEL		ALDEBARAN		* Mirfak	
90	48 03	043	27 37	104	41 41	147	23 37	169	31 50	194	53 02	216	64 32	287
91	48 30	043	28 15	105	42 02	148	23 45	170	31 40	195	52 38	217	63 55	287
92	48 57	043	28 53	106	42 23	149	23 51	171	31 30	196	52 14	219	63 17	288
93	49 24	044	29 31	107	42 42	151	23 58	172	31 18	197	51 49	220	62 40	288
94	49 52	044	30 08	107	43 01	152	24 03	173	31 06	198	51 23	221	62 02	288
95	50 19	044	30 46	108	43 20	153	24 07	174	30 54	199	50 57	223	61 25	289
96	50 46	044	31 23	109	43 37	154	24 11	175	30 40	201	50 30	224	60 48	289
97	51 14	044	32 00	110	43 54	156	24 15	176	30 26	202	50 02	225	60 10	289
98	51 41	044	32 37	111	44 10	157	24 17	177	30 11	203	49 34	227	59 33	290
99	52 09	045	33 14	112	44 25	158	24 19	178	29 55	204	49 05	228	58 56	290
100	52 36	045	33 50	113	44 39	160	24 20	179	29 39	205	48 35	229	58 19	291
101	53 04	045	34 27	114	44 52	161	24 20	180	29 22	206	48 05	231	57 43	291
102	53 32	045	35 02	115	45 04	163	24 20	181	29 05	207	47 35	232	57 06	291
103	54 00	045	35 38	116	45 16	164	24 18	182	28 46	208	47 03	233	56 29	292
104	54 28	045	36 14	117	45 26	165	24 17	183	28 27	209	46 32	234	55 53	292
	* Dubhe		Denebola		REGULUS		* SIRIUS		RIGEL		ALDEBARAN		* Mirfak	
105	54 56	045	22 58	094	36 49	118	24 14	184	28 08	210	46 00	235	55 16	293
106	55 24	045	23 38	095	37 23	119	24 11	185	27 47	211	45 27	236	54 40	293
107	55 52	046	24 17	095	37 58	120	24 07	186	27 27	212	44 54	237	54 04	293
108	56 20	046	24 56	096	38 32	121	24 02	187	27 05	214	44 21	239	53 28	294
109	56 48	046	25 35	097	39 06	122	23 56	188	26 43	215	43 47	240	52 52	294
110	57 16	046	26 14	098	39 39	123	23 50	190	26 21	216	43 13	241	52 16	295
111	57 45	046	26 53	099	40 12	124	23 43	191	25 57	217	42 38	242	51 40	295
112	58 13	046	27 32	099	40 45	125	23 36	192	25 34	218	42 03	243	51 04	295
113	58 41	046	28 11	100	41 17	126	23 28	193	25 09	219	41 28	244	50 29	296
114	59 09	046	28 50	101	41 49	127	23 19	194	24 44	220	40 53	245	49 53	296
115	59 37	046	29 28	102	42 20	128	23 09	195	24 19	221	40 17	246	49 18	297
116	60 06	046	30 07	103	42 51	129	22 59	196	23 53	222	39 41	247	48 43	297
117	60 34	046	30 45	104	43 21	130	22 48	197	23 27	223	39 05	248	48 08	297
118	61 02	046	31 23	104	43 51	131	22 36	198	23 00	223	38 28	249	47 33	298
119	61 30	046	32 01	105	44 20	133	22 24	199	22 33	224	37 51	250	46 58	298
	* Kochab		Denebola		* REGULUS		SIRIUS		RIGEL		* ALDEBARAN		CAPELLA	
120	43 26	021	32 39	106	44 49	134	22 11	200	22 05	225	37 14	251	62 10	280
121	43 40	021	33 17	107	45 17	135	21 57	201	21 37	226	36 37	251	61 31	280
122	43 55	022	33 55	108	45 45	136	21 43	202	21 08	227	36 00	252	60 52	281
123	44 09	022	34 32	109	46 12	138	21 28	203	20 39	228	35 22	253	60 13	281
124	44 24	022	35 09	110	46 38	139	21 12	204	20 09	229	34 44	254	59 35	282
125	44 39	022	35 46	111	47 04	140	20 56	205	19 39	230	34 06	255	58 56	282
126	44 54	022	36 23	112	47 29	141	20 39	206	19 09	231	33 28	256	58 18	283
127	45 09	022	36 59	112	47 53	143	20 22	207	18 38	232	32 50	257	57 40	283
128	45 24	023	37 36	113	48 16	144	20 04	208	18 07	233	32 11	258	57 01	284
129	45 39	023	38 12	114	48 39	145	19 45	209	17 36	234	31 33	258	56 23	285
130	45 54	023	38 48	115	49 01	147	19 26	210	17 04	234	30 54	259	55 45	285
131	46 10	023	39 23	116	49 22	148	19 06	211	16 32	235	30 16	260	55 07	286
132	46 25	023	39 58	117	49 43	150	18 46	212	15 59	236	29 37	261	54 29	286
133	46 41	023	40 33	118	50 02	151	18 25	212	15 26	237	28 58	262	53 51	287
134	46 56	023	41 08	119	50 21	152	18 04	213	14 53	238	28 19	263	53 14	287
	* Kochab		ARCTURUS		Denebola		* REGULUS		SIRIUS		BETELGEUSE		* CAPELLA	
135	47 12	023	21 56	085	41 42	120	50 39	154	17 42	214	32 56	239	52 36	288
136	47 27	024	22 35	086	42 15	121	50 56	155	17 19	215	32 22	240	51 59	288
137	47 43	024	23 14	087	42 49	123	51 12	157	16 56	216	31 47	241	51 21	289
138	47 59	024	23 54	088	43 22	124	51 27	158	16 33	217	31 13	242	50 44	289
139	48 15	024	24 33	088	43 55	125	51 41	160	16 09	218	30 38	243	50 07	290
140	48 31	024	25 12	089	44 27	126	51 54	161	15 45	219	30 02	244	49 30	290
141	48 47	024	25 52	090	44 58	127	52 06	163	15 20	220	29 27	245	48 53	290
142	49 03	024	26 31	091	45 30	128	52 17	165	14 54	221	28 51	246	48 16	291
143	49 19	024	27 10	091	46 00	129	52 27	166	14 28	222	28 15	247	47 39	291
144	49 35	024	27 50	092	46 31	131	52 36	168	14 02	222	27 39	248	47 03	292
145	49 51	024	28 29	093	47 00	132	52 44	169	13 35	223	27 02	248	46 26	292
146	50 07	024	29 08	094	47 30	133	52 51	171	13 08	224	26 26	249	45 50	293
147	50 23	024	29 48	094	47 58	134	52 56	173	12 40	225	25 49	250	45 14	293
148	50 40	024	30 27	095	48 26	135	53 01	174	12 12	226	25 12	251	44 37	294
149	50 56	024	31 06	096	48 53	137	53 04	176	11 44	227	24 34	252	44 02	294
	* Kochab		ARCTURUS		* SPICA		REGULUS		PROCYON		* BETELGEUSE		CAPELLA	
150	51 12	024	31 45	097	15 15	128	53 07	177	36 56	227	23 57	253	43 26	295
151	51 28	024	32 24	098	15 46	129	53 08	179	36 27	228	23 19	254	42 50	295
152	51 45	024	33 03	098	16 16	130	53 08	181	35 58	229	22 41	254	42 15	296
153	52 01	024	33 42	099	16 46	131	53 07	182	35 28	230	22 03	255	41 39	296
154	52 17	024	34 21	100	17 16	132	53 05	184	34 58	231	21 25	256	41 04	297
155	52 33	024	35 00	101	17 45	132	53 02	185	34 27	232	20 47	257	40 29	297
156	52 49	024	35 38	102	18 14	133	52 58	187	33 56	233	20 08	258	39 54	298
157	53 06	024	36 17	103	18 43	134	52 52	189	33 24	234	19 30	259	39 19	298
158	53 22	024	36 55	104	19 11	135	52 46	190	32 52	235	18 51	259	38 45	299
159	53 38	024	37 33	104	19 38	136	52 38	192	32 20	236	18 12	260	38 10	299
160	53 54	024	38 11	105	20 06	137	52 29	194	31 47	237	17 34	261	37 36	300
161	54 10	024	38 49	106	20 32	138	52 20	195	31 14	238	16 55	262	37 02	300
162	54 26	024	39 27	107	20 58	139	52 09	197	30 40	239	16 16	262	36 28	301
163	54 42	024	40 05	108	21 24	140	51 57	198	30 06	240	15 37	263	35 54	301
164	54 58	024	40 42	109	21 49	141	51 44	200	29 32	241	14 58	264	35 21	302
	* VEGA		ARCTURUS		* SPICA		REGULUS		PROCYON		* POLLUX		CAPELLA	
165	15 21	048	41 19	110	22 14	142	51 31	201	28 58	242	47 12	259	34 48	302
166	15 51	048	41 56	111	22 38	143	51 16	203	28 23	243	46 33	260	34 14	303
167	16 20	049	42 33	112	23 02	144	51 00	204	27 48	243	45 54	261	33 41	303
168	16 50	050	43 09	113	23 25	145	50 43	206	27 13	244	45 15	262	33 09	304
169	17 20	050	43 45	114	23 48	146	50 26	207	26 37	245	44 36	263	32 36	305
170	17 50	051	44 21	115	24 10	147	50 07	209	26 01	246	43 57	264	32 04	305
171	18 21	051	44 57	116	24 31	148	49 48	210	25 25	247	43 18	264	31 32	306
172	18 52	052	45 32	117	24 52	149	49 28	212	24 49	248	42 39	265	31 00	306
173	19 23	053	46 07	118	25 12	150	49 07	213	24 12	249	42 00	266	30 28	307
174	19 54	053	46 42	119	25 32	151	48 45	214	23 35	250	41 20	267	29 56	307
175	20 26	054	47 16	120	25 51	152	48 22	216	22 58	250	40 41	268	29 25	308
176	20 58	054	47 50	121	26 09	153	47 59	217	22 21	251	40 02	268	28 54	308
177	21 30	055	48 23	122	26 27	154	47 35	218	21 44	252	39 22	269	28 23	309
178	22 03	056	48 56	124	26 44	155	47 10	220	21 06	253	38 43	270	27 52	309
179	22 35	056	49 29	125	27 00	156	46 45	221	20 28	254	38 04	271	27 22	310

TABLE 5.—Precession and Nutation Correction

	Latitude															
LHA ♈	N89°	N80°	N70°	N60°	N50°	N40°	N20°	0°	S20°	S40°	S50°	S60°	S70°	S80°	S89°	LHA ♈
	mi. °	mi. °	mi. °	mi. °	mi. °	mi. °	mi. °	mi. °	mi. °	mi. °	mi. °	mi. °	mi. °	mi. °	mi. °	
								1966								
°																°
0	0 —	0 —	0 —	1 040	1 050	1 060	1 060	1 070	1 070	1 060	1 060	1 050	1 040	0 —	0 —	0
30	0 —	0 —	1 050	1 060	1 060	1 060	1 070	1 070	1 060	1 060	1 050	1 040	0 —	0 —	0 —	30
60	0 —	1 060	1 070	1 070	1 070	1 070	1 070	1 070	1 070	1 060	0 —	0 —	0 —	0 —	0 —	60
90	0 —	1 080	1 080	1 080	1 080	1 080	1 080	1 080	1 080	0 —	0 —	0 —	0 —	0 —	0 —	90
120	0 —	1 100	1 100	1 100	1 100	1 100	1 100	1 100	1 100	0 —	0 —	0 —	0 —	0 —	0 —	120
150	0 —	1 120	1 120	1 110	1 110	1 110	1 110	1 110	1 110	1 120	1 130	0 —	0 —	0 —	0 —	150
180	0 —	0 —	1 140	1 130	1 120	1 120	1 110	1 110	1 120	1 120	1 130	1 140	0 —	0 —	0 —	180
210	0 —	0 —	0 —	1 140	1 130	1 120	1 120	1 110	1 110	1 120	1 120	1 120	1 130	0 —	0 —	210
240	0 —	0 —	0 —	0 —	0 —	1 120	1 110	1 110	1 110	1 110	1 110	1 110	1 110	1 120	0 —	240
270	0 —	0 —	0 —	0 —	0 —	0 —	1 100	1 100	1 100	1 100	1 100	1 100	1 100	1 100	0 —	270
300	0 —	0 —	0 —	0 —	0 —	0 —	1 080	1 080	1 080	1 080	1 080	1 080	1 080	1 080	0 —	300
330	0 —	0 —	0 —	0 —	1 050	1 060	1 070	1 070	1 070	1 070	1 070	1 070	1 060	1 060	0 —	330
360	0 —	0 —	0 —	1 040	1 050	1 060	1 060	1 070	1 070	1 060	1 060	1 050	1 040	0 —	0 —	360
								1967								
0	1 350	1 010	1 030	1 050	1 050	1 060	2 060	2 070	2 070	2 060	1 060	1 050	1 040	1 030	1 010	0
30	1 020	1 040	1 050	1 060	2 060	2 060	2 070	2 070	2 070	1 060	1 050	1 040	1 030	1 000	1 340	30
60	1 050	1 060	1 070	1 070	2 070	2 070	2 080	2 070	2 070	1 060	1 050	1 040	0 —	1 330	1 310	60
90	1 080	1 080	1 080	2 090	2 090	2 090	2 090	2 090	1 090	1 080	1 080	0 —	0 —	0 —	1 280	90
120	1 110	1 100	1 100	2 100	2 100	2 100	2 100	2 100	1 100	1 110	1 120	0 —	0 —	0 —	1 250	120
150	1 140	1 130	1 120	1 110	2 110	2 110	2 110	2 110	2 110	1 120	1 130	1 140	1 170	1 200	1 220	150
180	1 170	1 150	1 140	1 130	1 120	2 120	2 110	2 110	2 120	1 120	1 130	1 130	1 150	1 170	1 190	180
210	1 200	1 180	1 150	1 140	1 130	1 120	2 110	2 110	2 110	2 120	2 120	1 120	1 130	1 140	1 160	210
240	1 230	1 210	0 —	1 140	1 130	1 120	2 110	2 110	2 100	2 110	2 110	1 110	1 110	1 120	1 130	240
270	1 260	0 —	0 —	0 —	1 100	1 100	1 090	2 090	2 090	2 090	2 090	2 090	1 100	1 100	1 100	270
300	1 290	0 —	0 —	0 —	1 060	1 070	1 080	2 080	2 080	2 080	2 080	2 080	1 080	1 080	1 070	300
330	1 320	1 340	1 010	1 040	1 050	1 060	2 070	2 070	2 070	2 070	2 070	1 070	1 060	1 050	1 040	330
360	1 350	1 010	1 030	1 050	1 050	1 060	2 060	2 070	2 070	2 060	1 060	1 050	1 040	1 030	1 010	360
								1968								
0	1 000	1 020	1 030	2 050	2 050	2 060	3 060	3 070	3 070	2 060	2 060	2 050	2 040	1 030	1 010	0
30	1 030	1 040	2 050	2 060	2 060	3 070	3 070	3 070	3 070	2 060	2 050	1 040	1 020	1 000	1 340	30
60	1 050	2 060	2 070	2 070	2 070	3 080	3 080	3 080	2 070	2 060	1 060	1 040	1 000	1 330	1 310	60
90	1 080	2 090	2 090	2 090	3 090	3 090	3 090	3 090	2 090	1 080	1 080	0 —	0 —	1 280	1 280	90
120	1 110	2 110	2 100	2 100	3 100	3 100	3 100	3 100	2 100	1 110	1 120	1 140	0 —	1 230	1 250	120
150	1 140	1 130	2 120	2 120	2 110	3 110	3 110	3 110	2 110	2 120	1 130	1 140	1 160	1 190	1 210	150
180	1 170	1 150	2 140	2 130	2 120	2 120	3 110	3 110	3 120	2 120	2 130	2 130	1 150	1 160	1 180	180
210	1 200	1 180	1 160	1 140	2 130	2 120	3 110	3 110	3 110	3 110	2 120	2 120	2 130	1 140	1 150	210
240	1 230	1 210	1 180	1 140	1 120	2 120	2 110	3 100	3 100	3 100	2 110	2 110	2 110	2 120	1 130	240
270	1 260	1 260	0 —	0 —	1 100	1 100	2 090	3 090	3 090	3 090	3 090	2 090	2 090	2 090	1 100	270
300	1 290	1 310	0 —	1 040	1 060	1 070	2 080	3 080	3 080	3 080	3 080	2 080	2 080	2 070	1 070	300
330	1 330	1 350	1 020	1 040	1 050	2 060	2 070	3 070	3 070	3 070	2 070	2 060	2 060	1 050	1 040	330
360	1 000	1 020	1 030	2 050	2 050	2 060	3 060	3 070	3 070	2 060	2 060	2 050	2 040	1 030	1 010	360
								1969								
0	2 000	2 020	2 040	2 050	3 050	3 060	4 070	4 070	4 070	3 060	3 060	2 050	2 040	2 030	2 010	0
30	2 030	2 040	2 050	3 060	3 060	3 070	4 070	4 070	3 070	3 060	2 050	2 040	1 020	1 000	1 340	30
60	2 060	2 060	3 070	3 070	3 070	4 080	4 080	4 080	3 070	2 060	2 060	1 040	1 000	1 330	1 310	60
90	2 080	2 090	3 090	3 090	3 090	4 090	4 090	3 090	3 090	2 090	1 080	0 —	0 —	1 280	1 280	90
120	2 110	2 110	3 100	3 100	3 100	4 100	4 100	4 100	3 100	2 110	1 120	1 140	1 190	1 230	1 240	120
150	2 140	2 130	2 120	3 120	3 110	3 110	4 110	4 110	3 110	2 120	2 130	2 140	1 160	1 190	1 210	150
180	2 170	2 150	2 140	2 130	3 120	3 120	4 110	4 110	4 110	3 120	3 130	2 130	2 140	2 160	2 180	180
210	1 200	1 180	1 160	2 140	2 130	3 120	3 110	4 110	4 110	3 110	3 120	3 120	2 130	2 140	2 150	210
240	1 230	1 210	1 180	1 140	2 120	2 120	3 110	4 100	4 100	4 100	3 110	3 110	3 110	2 120	2 120	240
270	1 260	1 260	0 —	0 —	1 100	2 090	3 090	3 090	4 090	4 090	3 090	3 090	3 090	2 090	2 100	270
300	1 300	1 310	1 350	1 040	1 060	2 070	3 080	4 080	4 080	4 080	3 080	3 080	3 080	2 070	2 070	300
330	1 330	1 350	1 020	2 040	2 050	2 060	3 070	4 070	4 070	3 070	3 070	3 060	2 060	2 050	2 040	330
360	2 000	2 020	2 040	2 050	3 050	3 060	4 070	4 070	4 070	3 060	3 060	2 050	2 040	2 030	2 010	360

N. Lat. {LHA greater than 180°....... Zn=Z; LHA less than 180°........... Zn=360−Z}

DECLINATION (0°-14°) CONTRARY NAME TO LATITUDE

LAT 38°

LHA	0°			1°			2°			3°			4°			5°			6°		
	Hc	d	Z	Hc	d	Z	Hc	d	Z	Hc	d	Z	Hc	d	Z	Hc	d	Z	Hc	d	Z
	° ′	′	°	° ′	′	°	° ′	′	°	° ′	′	°	° ′	′	°	° ′	′	°	° ′	′	°
69	16 24	38	103	15 46	39	104	15 07	39	105	14 28	39	106	13 49	39	107	13 10	40	107	12 30	39	108
68	17 10	39	104	16 31	38	105	15 53	40	106	15 13	39	106	14 34	39	107	13 55	40	108	13 15	40	109
67	17 56	39	105	17 17	39	105	16 38	39	106	15 59	40	107	15 19	39	108	14 40	40	109	14 00	40	109
66	18 42	39	105	18 03	40	106	17 23	39	107	16 44	40	108	16 04	40	109	15 24	40	109	14 44	40	110
65	19 27	−39	106	18 48	−40	107	18 08	−39	108	17 29	−40	108	16 49	−40	109	16 09	−40	110	15 29	−41	111
64	20 13	40	107	19 33	40	108	18 53	39	108	18 14	41	109	17 33	40	110	16 53	40	111	16 13	41	111
63	20 58	40	107	20 18	40	108	19 38	40	109	18 58	40	110	18 18	41	111	17 37	40	111	16 57	41	112
62	21 43	40	108	21 03	40	109	20 23	40	110	19 43	41	111	19 02	41	111	18 21	41	112	17 40	41	113
61	22 28	40	109	21 48	41	110	21 07	40	110	20 27	41	111	19 46	41	112	19 05	41	113	18 24	42	114
60	23 12	−40	110	22 32	−41	110	21 51	−40	111	21 11	−41	112	20 30	−42	113	19 48	−41	114	19 07	−42	114
59	23 57	41	110	23 16	41	111	22 35	41	112	21 54	41	113	21 13	41	114	20 32	42	114	19 50	42	115
58	24 41	41	111	24 00	41	112	23 19	41	113	22 38	42	113	21 56	41	114	21 15	42	115	20 33	42	116
57	25 25	41	112	24 44	41	113	24 03	42	113	23 21	42	114	22 39	42	115	21 57	42	116	21 15	42	117
56	26 09	42	113	25 27	41	113	24 46	42	114	24 04	42	115	23 22	42	116	22 40	43	117	21 57	42	117
55	26 52	−41	113	26 11	−42	114	25 29	−42	115	24 47	−42	116	24 05	−43	117	23 22	−43	117	22 39	−43	118
54	27 36	42	114	26 54	42	115	26 12	43	116	25 29	42	117	24 47	43	117	24 04	43	118	23 21	43	119
53	28 19	42	115	27 37	43	116	26 54	42	117	26 12	43	117	25 29	44	118	24 45	43	119	24 02	43	120
52	29 01	42	116	28 19	43	117	27 36	43	117	26 53	43	118	26 10	43	119	25 27	44	120	24 43	44	120
51	29 44	43	117	29 01	43	117	28 18	43	118	27 35	44	119	26 51	43	120	26 08	44	120	25 24	45	121
50	30 26	−43	117	29 43	−43	118	29 00	−44	119	28 16	−44	120	27 32	−44	121	26 48	−44	121	26 04	−45	122
49	31 08	43	118	30 25	44	119	29 41	44	120	28 57	44	121	28 13	44	121	27 29	45	122	26 44	45	123
48	31 49	43	119	31 06	44	120	30 22	44	121	29 38	45	121	28 53	45	122	28 08	45	123	27 23	45	124
47	32 31	44	120	31 47	45	121	31 02	44	122	30 18	45	122	29 33	45	123	28 48	45	124	28 03	46	125
46	33 11	44	121	32 27	45	122	31 42	44	122	30 58	46	123	30 12	45	124	29 27	46	125	28 41	46	125
45	33 52	−45	122	33 07	−45	122	32 22	−45	123	31 37	−46	124	30 51	−45	125	30 06	−46	126	29 20	−47	126
44	34 32	45	123	33 47	45	123	33 02	46	124	32 16	46	125	31 30	46	126	30 44	46	126	29 58	47	127
43	35 12	46	123	34 26	45	124	33 41	46	125	32 55	47	126	32 08	46	127	31 22	47	127	30 35	47	128
42	35 51	46	124	35 05	46	125	34 19	46	126	33 33	47	127	32 46	47	128	31 59	47	128	31 12	47	129
41	36 30	46	125	35 44	47	126	34 57	47	127	34 10	47	128	33 23	47	128	32 36	47	129	31 49	48	130
40	37 08	−47	126	36 21	−46	127	35 35	−47	128	34 48	−48	129	34 00	−47	129	33 13	−48	130	32 25	−48	131
39	37 46	47	127	36 59	47	128	36 12	48	129	35 24	47	130	34 37	48	130	33 49	49	131	33 00	48	132
38	38 23	47	128	37 36	48	129	36 48	48	130	36 00	48	131	35 12	48	131	34 24	49	132	33 35	49	133
37	39 00	48	129	38 12	48	130	37 24	48	131	36 36	48	132	35 48	49	132	34 59	49	133	34 10	50	134
36	39 36	48	130	38 48	48	131	38 00	49	132	37 11	49	133	36 22	49	133	35 33	49	134	34 44	50	135
35	40 12	−48	131	39 24	−49	132	38 35	−49	133	37 46	−50	134	36 56	−49	134	36 07	−50	135	35 17	−50	136
34	40 47	48	132	39 59	50	133	39 09	49	134	38 20	50	135	37 30	50	135	36 40	50	136	35 50	51	137
33	41 22	49	134	40 33	50	134	39 43	50	135	38 53	50	136	38 03	50	136	37 13	51	137	36 22	51	138
32	41 56	50	135	41 06	50	135	40 16	50	136	39 26	51	137	38 35	51	137	37 44	51	138	36 53	51	139
31	42 29	50	136	41 39	50	136	40 49	51	137	39 58	51	138	39 07	51	139	38 16	52	139	37 24	52	140
30	43 02	−51	137	42 11	−51	138	41 20	−51	138	40 29	−51	139	39 38	−52	140	38 46	−52	140	37 54	−52	141
29	43 34	51	138	42 43	51	139	41 52	52	139	41 00	52	140	40 08	52	141	39 16	52	141	38 24	53	142
28	44 05	51	139	43 14	52	140	42 22	52	141	41 30	52	141	40 38	53	142	39 45	53	143	38 52	52	143
27	44 36	52	140	43 44	52	141	42 52	53	142	41 59	53	142	41 06	52	143	40 14	54	144	39 20	53	144
26	45 06	53	142	44 13	52	142	43 21	53	143	42 28	54	144	41 34	53	144	40 41	53	145	39 48	54	145
25	45 35	−53	143	44 42	−53	144	43 49	−54	144	42 55	−53	145	42 02	−54	145	41 08	−54	146	40 14	−54	147
24	46 03	54	144	45 09	53	145	44 16	54	145	43 22	54	146	42 28	54	147	41 34	54	147	40 40	55	148
23	46 30	54	145	45 36	54	146	44 42	54	147	43 48	54	147	42 54	55	148	41 59	54	148	41 05	55	149
22	46 56	54	147	46 02	54	147	45 08	55	148	44 13	55	149	43 18	54	149	42 24	56	150	41 28	55	150
21	47 22	55	148	46 27	55	149	45 32	55	149	44 37	55	150	43 42	55	150	42 47	55	151	41 52	56	151
20	47 46	−55	149	46 51	−55	150	45 56	−55	151	45 01	−56	151	44 05	−55	152	43 10	−56	152	42 14	−56	153
19	48 10	55	151	47 15	56	151	46 19	56	152	45 23	56	152	44 27	56	153	43 31	56	153	42 35	56	154
18	48 33	56	152	47 37	56	153	46 41	56	153	45 45	57	154	44 48	56	154	43 52	57	155	42 55	56	155
17	48 54	56	154	47 58	57	154	47 01	56	155	46 05	57	155	45 08	56	156	44 12	57	156	43 15	57	157
16	49 15	57	155	48 18	57	156	47 21	57	156	46 24	57	157	45 27	57	157	44 30	57	157	43 33	57	158
15	49 34	−57	157	48 37	−57	157	47 40	−57	157	46 43	−58	158	45 45	−57	158	44 48	−58	159	43 50	−57	159
14	49 52	57	158	48 55	57	158	47 58	58	159	47 00	58	159	46 02	57	160	45 05	58	160	44 07	58	160
13	50 10	58	159	49 12	58	160	48 14	58	160	47 16	58	161	46 18	58	161	45 20	58	161	44 22	58	162
12	50 26	58	161	49 28	59	161	48 29	58	162	47 31	58	162	46 33	58	162	45 35	59	163	44 36	58	163
11	50 40	58	163	49 42	58	163	48 44	59	163	47 45	58	164	46 47	59	164	45 48	58	164	44 50	59	165
10	50 54	−59	164	49 55	−58	164	48 57	−59	165	47 58	−59	165	46 59	−59	165	46 00	−58	166	45 02	−59	166
9	51 06	58	166	50 08	59	166	49 09	59	166	48 10	59	167	47 11	59	167	46 12	59	167	45 13	59	167
8	51 18	60	167	50 18	59	167	49 19	59	168	48 20	59	168	47 21	59	168	46 22	59	168	45 23	60	169
7	51 27	59	169	50 28	59	169	49 29	60	169	48 29	59	169	47 30	59	170	46 31	60	170	45 31	59	170
6	51 36	59	170	50 37	60	171	49 37	59	171	48 38	60	171	47 38	60	171	46 38	59	171	45 39	60	171
5	51 43	−59	172	50 44	−60	172	49 44	−60	172	48 44	−59	172	47 45	−60	173	46 45	−60	173	45 45	−59	173
4	51 49	59	174	50 50	60	174	49 50	60	174	48 50	60	174	47 50	60	174	46 50	59	174	45 51	60	174
3	51 54	60	175	50 54	60	175	49 54	60	175	48 54	59	175	47 55	60	176	46 55	60	176	45 55	60	176
2	51 57	60	177	50 57	60	177	49 57	59	177	48 58	60	177	47 58	60	177	46 58	60	177	45 58	60	177
1	51 59	60	178	50 59	60	178	49 59	60	178	48 59	60	179	47 59	60	179	46 59	60	179	45 59	60	179
0	52 00	−60	180	51 00	−60	180	50 00	−60	180	49 00	−60	180	48 00	−60	180	47 00	−60	180	46 00	−60	180
	0°			1°			2°			3°			4°			5°			6°		

LHA	7°			8°			9°			10°			11°			12°			13°			14°			LHA
	Hc	d	Z	Hc	d	Z	Hc	d	Z	Hc	d	Z	Hc	d	Z	Hc	d	Z	Hc	d	Z	Hc	d	Z	
	° ′	′	°	° ′	′	°	° ′	′	°	° ′	′	°	° ′	′	°	° ′	′	°	° ′	′	°	° ′	′	°	
69	11 51	40	109	11 11	40	110	10 31	39	110	09 52	40	111	09 12	41	112	08 31	40	113	07 51	40	113	07 11	40	114	291
68	12 35	39	109	11 56	40	110	11 16	40	111	10 36	41	112	09 55	40	113	09 15	40	113	08 35	41	114	07 54	40	115	292
67	13 20	40	110	12 40	40	111	12 00	41	112	11 19	40	112	10 39	41	113	09 58	40	114	09 18	41	115	08 37	41	115	293
66	14 04	40	111	13 24	41	112	12 43	40	112	12 03	41	113	11 22	41	114	10 41	40	115	10 01	41	115	09 20	41	116	294
65	14 48	−40	112	14 08	−41	112	13 27	−41	113	12 46	−41	114	12 05	−41	115	11 24	−41	115	10 43	−41	116	10 02	−41	117	295
64	15 32	41	112	14 51	40	113	14 11	42	114	13 29	41	115	12 48	41	115	12 07	41	116	11 26	42	117	10 44	42	117	296
63	16 16	41	113	15 35	41	114	14 54	42	114	14 12	41	115	13 31	42	116	12 49	41	117	12 08	42	117	11 26	42	118	297
62	16 59	41	114	16 18	41	114	15 37	42	115	14 55	42	116	14 13	41	117	13 32	42	117	12 50	43	118	12 07	42	119	298
61	17 42	41	114	17 01	42	115	16 19	42	116	15 37	41	117	14 56	43	117	14 13	42	118	13 31	42	119	12 49	43	120	299
60	18 25	−41	115	17 44	−42	116	17 02	−42	117	16 20	−43	117	15 37	−42	118	14 55	−43	119	14 12	−42	120	13 30	−43	120	300
59	19 08	42	116	18 26	42	117	17 44	42	117	17 02	43	118	16 19	43	119	15 36	43	120	14 53	43	120	14 10	43	121	301
58	19 51	43	117	19 08	42	117	18 26	43	118	17 43	43	119	17 00	43	120	16 17	43	120	15 34	43	121	14 51	44	122	302
57	20 33	43	117	19 50	43	118	19 07	43	119	18 24	43	120	17 41	43	120	16 58	43	121	16 15	44	122	15 31	44	122	303
56	21 15	43	118	20 32	43	119	19 49	44	120	19 05	43	120	18 22	44	121	17 38	43	122	16 55	44	122	16 11	44	123	304
55	21 56	−43	119	21 13	−43	120	20 30	−44	120	19 46	−44	121	19 02	−44	122	18 18	−44	122	17 34	−44	123	16 50	−44	124	305
54	22 38	44	120	21 54	44	120	21 10	44	121	20 26	44	122	19 42	44	123	18 58	44	123	18 14	45	124	17 29	44	125	306
53	23 19	44	120	22 35	44	121	21 51	44	122	21 07	45	123	20 22	44	123	19 38	45	124	18 53	45	125	18 08	45	125	307
52	23 59	44	121	23 15	44	122	22 31	45	123	21 46	45	123	21 01	44	124	20 17	45	125	19 32	46	125	18 46	45	126	308
51	24 39	44	122	23 55	45	123	23 10	44	123	22 26	46	124	21 40	45	125	20 55	45	126	20 10	46	126	19 24	45	127	309
50	25 19	−44	123	24 35	−45	124	23 50	−46	124	23 04	−45	125	22 19	−45	126	21 34	−46	126	20 48	−46	127	20 02	−46	128	310
49	25 59	45	124	25 14	45	124	24 29	46	125	23 43	46	126	22 57	46	126	22 11	46	127	21 25	46	128	20 39	46	129	311
48	26 38	45	124	25 53	46	125	25 07	46	126	24 21	46	127	23 35	46	127	22 49	46	128	22 03	47	129	21 16	47	129	312
47	27 17	46	125	26 31	46	126	25 45	46	127	24 59	46	127	24 13	47	128	23 26	47	129	22 39	47	129	21 52	47	130	313
46	27 55	46	126	27 09	46	127	26 23	47	128	25 36	46	128	24 50	47	129	24 03	47	130	23 16	48	130	22 28	47	131	314
45	28 33	−46	127	27 47	−47	128	27 00	−47	128	26 13	−47	129	25 26	−47	130	24 39	−48	130	23 51	−47	131	23 04	−48	132	315
44	29 11	47	128	28 24	47	129	27 37	47	129	26 50	48	130	26 02	47	131	25 15	48	131	24 27	48	132	23 39	48	133	316
43	29 48	47	129	29 01	48	129	28 13	47	130	27 26	48	131	26 38	48	132	25 50	48	132	25 02	49	133	24 13	48	134	317
42	30 25	48	130	29 37	48	130	28 49	48	131	28 01	48	132	27 13	48	132	26 25	49	133	25 36	49	134	24 47	49	134	318
41	31 01	48	131	30 13	48	131	29 25	49	132	28 36	48	133	27 48	49	133	26 59	49	134	26 10	49	135	25 21	49	135	319
40	31 37	−49	132	30 48	−48	132	30 00	−49	133	29 11	−49	134	28 22	−49	134	27 33	−50	135	26 43	−49	136	25 54	−50	136	320
39	32 12	49	132	31 23	49	133	30 34	49	134	29 45	49	135	28 56	50	135	28 06	50	136	27 16	50	136	26 26	50	137	321
38	32 46	49	133	31 57	49	134	31 08	50	135	30 18	49	135	29 29	50	136	28 39	50	137	27 49	51	137	26 58	50	138	322
37	33 20	49	134	32 31	50	135	31 41	50	136	30 51	50	136	30 01	50	137	29 11	51	138	28 20	50	138	27 30	51	139	323
36	33 54	50	136	33 04	50	136	32 14	50	137	31 24	51	137	30 33	50	138	29 43	51	139	28 52	51	139	28 01	51	140	324
35	34 27	−50	136	33 37	−51	137	32 46	−51	138	31 55	−50	138	31 05	−51	139	30 14	−52	140	29 22	−51	140	28 31	−52	141	325
34	34 59	50	137	34 09	51	138	33 18	51	139	32 27	52	139	31 35	51	140	30 44	52	141	29 52	51	141	29 01	52	142	326
33	35 31	51	138	34 40	51	139	33 49	52	140	32 57	51	140	32 06	52	141	31 14	52	142	30 22	52	142	29 30	53	143	327
32	36 02	51	139	35 11	52	140	34 19	52	141	33 27	52	141	32 35	52	142	31 43	53	143	30 50	52	143	29 58	53	144	328
31	36 32	51	141	35 41	52	141	34 49	53	142	33 56	52	142	33 04	53	143	32 11	52	144	31 19	53	144	30 26	53	145	329
30	37 02	−52	142	36 10	−53	142	35 17	−52	143	34 25	−53	143	33 32	−53	144	32 39	−53	145	31 46	−53	145	30 53	−54	146	330
29	37 31	52	143	36 39	53	143	35 46	53	144	34 53	53	144	34 00	54	145	33 06	53	146	32 13	54	146	31 19	54	147	331
28	38 00	53	144	37 07	54	144	36 13	53	145	35 20	54	146	34 26	53	146	33 33	54	147	32 39	54	147	31 45	54	148	332
27	38 27	53	145	37 34	54	145	36 40	54	146	35 46	54	147	34 52	54	147	33 58	54	148	33 04	54	148	32 10	55	149	333
26	38 54	54	146	38 00	54	147	37 06	54	147	36 12	54	148	35 18	55	148	34 23	54	149	33 29	55	149	32 34	55	150	334
25	39 20	−54	147	38 26	−55	148	37 31	−54	148	36 37	−55	149	35 42	−55	149	34 47	−54	150	33 53	−55	150	32 58	−56	151	335
24	39 45	54	148	38 51	55	149	37 56	55	149	37 01	55	150	36 06	55	150	35 11	55	151	34 16	56	151	33 20	55	152	336
23	40 10	55	150	39 15	55	150	38 20	56	151	37 24	55	151	36 29	56	152	35 33	55	152	34 38	56	152	33 42	56	153	337
22	40 33	55	151	39 38	56	151	38 42	55	152	37 47	56	152	36 51	56	153	35 55	56	153	34 59	56	154	34 03	56	154	338
21	40 56	56	152	40 00	56	152	39 04	56	153	38 08	56	153	37 12	56	154	36 16	56	154	35 20	56	155	34 24	57	155	339
20	41 18	−56	153	40 22	−56	154	39 26	−57	154	38 29	−56	155	37 33	−57	155	36 36	−56	155	35 40	−57	156	34 43	−57	156	340
19	41 39	57	154	40 42	56	155	39 46	57	155	38 49	57	156	37 52	56	156	36 56	57	157	35 59	57	157	35 02	57	157	341
18	41 59	57	156	41 02	57	156	40 05	57	157	39 08	57	157	38 11	57	157	37 14	57	158	36 17	57	158	35 20	58	158	342
17	42 18	57	157	41 21	58	157	40 23	57	158	39 26	57	158	38 29	57	159	37 32	58	159	36 34	57	159	35 37	58	160	343
16	42 36	58	158	41 38	57	159	40 41	58	159	39 43	57	159	38 46	58	160	37 48	58	160	36 50	57	160	35 53	58	161	344
15	42 53	−58	160	41 55	−58	160	40 57	−57	160	40 00	−58	161	39 02	−58	161	38 04	−58	161	37 06	−58	162	36 08	−58	162	345
14	43 09	58	161	42 11	58	161	41 13	58	162	40 15	58	162	39 17	58	162	38 19	59	162	37 20	58	163	36 22	58	163	346
13	43 24	58	162	42 26	59	162	41 27	58	163	40 29	58	163	39 31	59	163	38 32	58	164	37 34	59	164	36 35	58	164	347
12	43 38	59	163	42 39	58	164	41 41	59	164	40 42	58	164	39 44	59	165	38 45	59	165	37 46	58	165	36 48	59	165	348
11	43 51	59	165	42 52	59	165	41 53	58	165	40 55	59	166	39 56	59	166	38 57	59	166	37 58	59	166	36 59	59	167	349
10	44 03	−59	166	43 04	−59	166	42 05	−59	167	41 06	−59	167	40 07	−59	167	39 08	−59	167	38 09	−59	168	37 10	−59	168	350
9	44 14	60	168	43 14	59	168	42 15	59	168	41 16	59	168	40 17	59	168	39 18	60	169	38 18	59	169	37 19	59	169	351
8	44 23	59	169	43 24	59	169	42 25	60	169	41 25	59	170	40 26	59	170	39 27	60	170	38 27	59	170	37 28	60	170	352
7	44 32	60	170	43 32	59	170	42 33	60	171	41 33	59	171	40 34	60	171	39 34	59	171	38 35	60	171	37 35	59	171	353
6	44 39	59	172	43 40	60	172	42 40	60	172	41 40	59	172	40 41	60	172	39 41	60	172	38 41	59	173	37 42	60	173	354
5	44 46	−60	173	43 46	−60	173	42 46	−60	173	41 46	−59	173	40 47	−60	174	39 47	−60	174	38 47	−60	174	37 47	−59	174	355
4	44 51	60	174	43 51	60	175	42 51	60	175	41 51	60	175	40 51	59	175	39 52	60	175	38 52	60	175	37 52	60	175	356
3	44 55	60	176	43 55	60	176	42 55	60	176	41 55	60	176	40 55	60	176	39 55	60	176	38 55	60	176	37 55	59	176	357
2	44 58	60	177	43 58	60	177	42 58	60	177	41 58	60	177	40 58	60	177	39 58	60	177	38 58	60	178	37 58	60	178	358
1	44 59	60	179	43 59	60	179	42 59	59	179	42 00	60	179	41 00	60	179	40 00	60	179	39 00	60	179	38 00	50	179	359
0	45 00	−60	180	44 00	−60	180	43 00	−60	180	42 00	−60	180	41 00	−60	180	40 00	−60	180	39 00	−60	180	38 00	−60	180	360
	7°			8°			9°			10°			11°			12°			13°			14°			

S. Lat. {LHA greater than 180°....... Zn=180−Z; LHA less than 180°........... Zn=180+Z}

DECLINATION (0°-14°) CONTRARY NAME TO LATITUDE

LAT 38°

TABLE 4.—Correction to Tabulated Altitude for Minutes of Declination

d / ′	1	2	3	4	5	6	7	8	9	10	11	12	13	14	15	16	17	18	19	20	21	22	23	24	25	26	27	28	29	30	31	32	33	34	35	36	37	38	39	40	41	42	43	44	45	46	47	48	49	50	51	52	53	54	55	56	57	58	59	60	d / ′
0	0	0	0	0	0	0	0	0	0	0	0	0	0	0	0	0	0	0	0	0	0	0	0	0	0	0	0	0	0	0	0	0	0	0	0	0	0	0	0	0	0	0	0	0	0	0	0	0	0	0	0	0	0	0	0	0	0	0	0	0	0
1	0	0	0	0	0	0	0	0	0	0	0	0	0	0	0	0	0	0	0	0	0	0	0	0	0	0	0	0	0	0	1	1	1	1	1	1	1	1	1	1	1	1	1	1	1	1	1	1	1	1	1	1	1	1	1	1	1	1	1	1	1
2	0	0	0	0	0	0	0	0	0	0	0	0	0	0	0	1	1	1	1	1	1	1	1	1	1	1	1	1	1	1	1	1	1	1	1	1	1	1	1	1	1	1	1	1	2	2	2	2	2	2	2	2	2	2	2	2	2	2	2	2	2
3	0	0	0	0	0	0	0	0	0	0	1	1	1	1	1	1	1	1	1	1	1	1	1	1	1	1	1	1	1	2	2	2	2	2	2	2	2	2	2	2	2	2	2	2	2	2	2	2	2	2	3	3	3	3	3	3	3	3	3	3	3
4	0	0	0	0	0	0	0	1	1	1	1	1	1	1	1	1	1	1	1	1	1	1	2	2	2	2	2	2	2	2	2	2	2	2	2	2	2	3	3	3	3	3	3	3	3	3	3	3	3	3	3	3	4	4	4	4	4	4	4	4	4
5	0	0	0	0	0	0	1	1	1	1	1	1	1	1	1	1	1	2	2	2	2	2	2	2	2	2	2	2	2	2	3	3	3	3	3	3	3	3	3	3	3	4	4	4	4	4	4	4	4	4	4	4	4	4	5	5	5	5	5	5	5
6	0	0	0	0	0	1	1	1	1	1	1	1	1	1	2	2	2	2	2	2	2	2	2	2	2	3	3	3	3	3	3	3	3	3	4	4	4	4	4	4	4	4	4	4	4	5	5	5	5	5	5	5	5	5	6	6	6	6	6	6	6
7	0	0	0	0	1	1	1	1	1	1	1	1	2	2	2	2	2	2	2	2	2	3	3	3	3	3	3	3	3	4	4	4	4	4	4	4	4	4	5	5	5	5	5	5	5	5	5	6	6	6	6	6	6	6	6	7	7	7	7	7	7
8	0	0	0	1	1	1	1	1	1	1	1	2	2	2	2	2	2	2	3	3	3	3	3	3	3	3	4	4	4	4	4	4	4	5	5	5	5	5	5	5	5	6	6	6	6	6	6	6	7	7	7	7	7	7	7	7	8	8	8	8	8
9	0	0	0	1	1	1	1	1	1	2	2	2	2	2	2	2	3	3	3	3	3	3	3	4	4	4	4	4	4	4	5	5	5	5	5	5	6	6	6	6	6	6	6	7	7	7	7	7	7	8	8	8	8	8	8	8	9	9	9	9	9
10	0	0	0	1	1	1	1	1	2	2	2	2	2	2	2	3	3	3	3	3	4	4	4	4	4	4	4	5	5	5	5	5	6	6	6	6	6	6	6	7	7	7	7	7	8	8	8	8	8	8	8	9	9	9	9	9	10	10	10	10	10
11	0	0	1	1	1	1	1	1	2	2	2	2	2	3	3	3	3	3	3	4	4	4	4	4	5	5	5	5	5	6	6	6	6	6	6	7	7	7	7	7	8	8	8	8	8	8	9	9	9	9	9	10	10	10	10	10	10	11	11	11	11
12	0	0	1	1	1	1	1	2	2	2	2	2	3	3	3	3	3	4	4	4	4	4	5	5	5	5	5	6	6	6	6	6	7	7	7	7	7	8	8	8	8	8	9	9	9	9	9	10	10	10	10	10	11	11	11	11	11	12	12	12	12
13	0	0	1	1	1	1	2	2	2	2	2	3	3	3	3	3	4	4	4	4	5	5	5	5	5	6	6	6	6	6	7	7	7	7	8	8	8	8	8	9	9	9	9	10	10	10	10	10	11	11	11	11	11	12	12	12	12	13	13	13	13
14	0	0	1	1	1	1	2	2	2	2	3	3	3	3	4	4	4	4	4	5	5	5	5	6	6	6	6	7	7	7	7	7	8	8	8	8	9	9	9	9	10	10	10	10	10	11	11	11	11	12	12	12	12	13	13	13	13	14	14	14	14
15	0	0	1	1	1	2	2	2	2	2	3	3	3	4	4	4	4	4	5	5	5	6	6	6	6	6	7	7	7	8	8	8	8	8	9	9	9	10	10	10	10	10	11	11	11	12	12	12	12	12	13	13	13	14	14	14	14	14	15	15	15
16	0	1	1	1	1	2	2	2	2	3	3	3	3	4	4	4	5	5	5	5	6	6	6	6	7	7	7	7	8	8	8	9	9	9	9	10	10	10	10	11	11	11	11	12	12	12	13	13	13	13	14	14	14	14	15	15	15	15	16	16	16
17	0	1	1	1	1	2	2	2	3	3	3	3	4	4	4	5	5	5	5	6	6	6	7	7	7	7	8	8	8	8	9	9	9	10	10	10	10	11	11	11	12	12	12	12	13	13	13	14	14	14	14	15	15	15	16	16	16	16	17	17	17
18	0	1	1	1	2	2	2	2	3	3	3	4	4	4	4	5	5	5	6	6	6	7	7	7	8	8	8	8	9	9	9	10	10	10	10	11	11	11	12	12	12	13	13	13	14	14	14	14	15	15	15	16	16	16	16	17	17	17	18	18	18
19	0	1	1	1	2	2	2	3	3	3	3	4	4	4	5	5	5	6	6	6	7	7	7	8	8	8	9	9	9	10	10	10	10	11	11	11	12	12	12	13	13	13	14	14	14	15	15	15	16	16	16	16	17	17	17	18	18	18	19	19	19
20	0	1	1	1	2	2	2	3	3	3	4	4	4	5	5	5	6	6	6	7	7	7	8	8	8	9	9	9	10	10	10	11	11	11	12	12	12	13	13	13	14	14	14	15	15	15	16	16	16	17	17	17	18	18	18	19	19	19	20	20	20
21	0	1	1	1	2	2	2	3	3	4	4	4	5	5	5	6	6	6	7	7	7	8	8	8	9	9	9	10	10	10	11	11	12	12	12	13	13	13	14	14	14	15	15	15	16	16	16	17	17	18	18	18	19	19	19	20	20	20	21	21	21
22	0	1	1	1	2	2	3	3	3	4	4	4	5	5	6	6	6	7	7	7	8	8	8	9	9	10	10	10	11	11	11	12	12	12	13	13	14	14	14	15	15	15	16	16	16	17	17	18	18	18	19	19	19	20	20	21	21	21	22	22	22
23	0	1	1	2	2	2	3	3	3	4	4	5	5	5	6	6	7	7	7	8	8	8	9	9	10	10	10	11	11	12	12	12	13	13	13	14	14	15	15	15	16	16	16	17	17	18	18	18	19	19	20	20	20	21	21	21	22	22	23	23	23
24	0	1	1	2	2	2	3	3	4	4	4	5	5	6	6	6	7	7	8	8	8	9	9	10	10	10	11	11	12	12	12	13	13	14	14	14	15	15	16	16	16	17	17	18	18	18	19	19	20	20	20	21	21	22	22	22	23	23	24	24	24
25	0	1	1	2	2	2	3	3	4	4	5	5	5	6	6	7	7	8	8	8	9	9	10	10	10	11	11	12	12	12	13	13	14	14	15	15	15	16	16	17	17	18	18	18	19	19	20	20	20	21	21	22	22	22	23	23	24	24	25	25	25
26	0	1	1	2	2	3	3	3	4	4	5	5	6	6	6	7	7	8	8	9	9	10	10	10	11	11	12	12	13	13	13	14	14	15	15	16	16	16	17	17	18	18	19	19	20	20	20	21	21	22	22	23	23	23	24	24	25	25	26	26	26
27	0	1	1	2	2	3	3	4	4	4	5	5	6	6	7	7	8	8	9	9	9	10	10	11	11	12	12	13	13	14	14	14	15	15	16	16	17	17	18	18	18	19	19	20	20	21	21	22	22	22	23	23	24	24	25	25	26	26	27	27	27
28	0	1	1	2	2	3	3	4	4	5	5	6	6	7	7	7	8	8	9	9	10	10	11	11	12	12	13	13	14	14	14	15	15	16	16	17	17	18	18	19	19	20	20	21	21	21	22	22	23	23	24	24	25	25	26	26	27	27	28	28	28
29	0	1	1	2	2	3	3	4	4	5	5	6	6	7	7	8	8	9	9	10	10	11	11	12	12	13	13	14	14	14	15	15	16	16	17	17	18	18	19	19	20	20	21	21	22	22	23	23	24	24	25	25	26	26	27	27	28	28	29	29	29
30	0	1	2	2	2	3	4	4	4	5	6	6	6	7	8	8	8	9	10	10	10	11	12	12	12	13	14	14	14	15	16	16	16	17	18	18	18	19	20	20	20	21	22	22	22	23	24	24	24	25	26	26	26	27	28	28	28	29	30	30	30
31	1	1	2	2	3	3	4	4	5	5	6	6	7	7	8	8	9	9	10	10	11	11	12	12	13	13	14	14	15	16	16	17	17	18	18	19	19	20	20	21	21	22	22	23	23	24	24	25	25	26	26	27	27	28	28	29	29	30	30	31	31
32	1	1	2	2	3	3	4	4	5	5	6	6	7	7	8	9	9	10	10	11	11	12	12	13	13	14	14	15	15	16	17	17	18	18	19	19	20	20	21	21	22	22	23	23	24	25	25	26	26	27	27	28	28	29	29	30	30	31	31	32	32
33	1	1	2	2	3	3	4	4	5	6	6	7	7	8	8	9	9	10	10	11	12	12	13	13	14	14	15	15	16	16	17	18	18	19	19	20	20	21	21	22	23	23	24	24	25	25	26	26	27	28	28	29	29	30	30	31	31	32	32	33	33
34	1	1	2	2	3	3	4	5	5	6	6	7	7	8	8	9	10	10	11	11	12	12	13	14	14	15	15	16	16	17	18	18	19	19	20	20	21	22	22	23	23	24	24	25	26	26	27	27	28	28	29	29	30	31	31	32	32	33	33	34	34
35	1	1	2	2	3	4	4	5	5	6	6	7	8	8	9	9	10	10	11	12	12	13	13	14	15	15	16	16	17	18	18	19	19	20	20	21	22	22	23	23	24	24	25	26	26	27	27	28	29	29	30	30	31	32	32	33	33	34	34	35	35
36	1	1	2	2	3	4	4	5	5	6	7	7	8	8	9	10	10	11	11	12	13	13	14	14	15	16	16	17	17	18	19	19	20	20	21	22	22	23	23	24	25	25	26	26	27	28	28	29	29	30	31	31	32	32	33	34	34	35	35	36	36
37	1	1	2	2	3	4	4	5	6	6	7	7	8	9	9	10	10	11	12	12	13	14	14	15	15	16	17	17	18	18	19	20	20	21	22	22	23	23	24	25	25	26	27	27	28	28	29	30	30	31	31	32	33	33	34	35	35	36	36	37	37
38	1	1	2	3	3	4	4	5	6	6	7	8	8	9	10	10	11	11	12	13	13	14	15	15	16	16	17	18	18	19	20	20	21	22	22	23	23	24	25	25	26	27	27	28	28	29	30	30	31	32	32	33	34	34	35	35	36	37	37	38	38
39	1	1	2	3	3	4	5	5	6	6	7	8	8	9	10	10	11	12	12	13	14	14	15	16	16	17	18	18	19	20	20	21	21	22	23	23	24	25	25	26	27	27	28	29	29	30	31	31	32	32	33	34	34	35	36	36	37	38	38	39	39
40	1	1	2	3	3	4	5	5	6	7	7	8	9	9	10	11	11	12	13	13	14	15	15	16	17	17	18	19	19	20	21	21	22	23	23	24	25	25	26	27	27	28	29	29	30	31	31	32	33	33	34	35	35	36	37	37	38	39	39	40	40
41	1	1	2	3	3	4	5	5	6	7	8	8	9	10	10	11	12	12	13	14	14	15	16	16	17	18	18	19	20	20	21	22	23	23	24	25	25	26	27	27	28	29	29	30	31	31	32	33	33	34	35	36	36	37	38	38	39	40	40	41	41
42	1	1	2	3	4	4	5	6	6	7	8	8	9	10	10	11	12	13	13	14	15	15	16	17	18	18	19	20	20	21	22	22	23	24	24	25	26	27	27	28	29	29	30	31	32	32	33	34	34	35	36	36	37	38	38	39	40	41	41	42	42
43	1	1	2	3	4	4	5	6	6	7	8	9	9	10	11	11	12	13	14	14	15	16	16	17	18	19	19	20	21	22	22	23	24	24	25	26	27	27	28	29	29	30	31	32	32	33	34	34	35	36	37	37	38	39	39	40	41	42	42	43	43
44	1	1	2	3	4	4	5	6	7	7	8	9	10	10	11	12	12	13	14	15	15	16	17	18	18	19	20	21	21	22	23	23	24	25	26	26	27	28	29	29	30	31	32	32	33	34	34	35	36	37	37	38	39	40	40	41	42	43	43	44	44
45	1	2	2	3	4	4	5	6	7	8	8	9	10	10	11	12	13	14	14	15	16	16	17	18	19	20	20	21	22	22	23	24	25	26	26	27	28	28	29	30	31	32	32	33	34	34	35	36	37	38	38	39	40	40	41	42	43	44	44	45	45
46	1	2	2	3	4	5	5	6	7	8	8	9	10	11	12	12	13	14	15	15	16	17	18	18	19	20	21	21	22	23	24	25	25	26	27	28	28	29	30	31	31	32	33	34	34	35	36	37	38	38	39	40	41	41	42	43	44	44	45	46	46
47	1	2	2	3	4	5	5	6	7	8	9	9	10	11	12	13	13	14	15	16	16	17	18	19	20	20	21	22	23	24	24	25	26	27	27	28	29	30	31	31	32	33	34	34	35	36	37	38	38	39	40	41	42	42	43	44	45	45	46	47	47
48	1	2	2	3	4	5	6	6	7	8	9	10	10	11	12	13	14	14	15	16	17	18	18	19	20	21	22	22	23	24	25	26	26	27	28	29	30	30	31	32	33	34	34	35	36	37	38	38	39	40	41	42	42	43	44	45	46	46	47	48	48
49	1	2	2	3	4	5	6	7	7	8	9	10	11	11	12	13	14	15	16	16	17	18	19	20	20	21	22	23	24	24	25	26	27	28	29	29	30	31	32	33	33	34	35	36	37	38	38	39	40	41	42	42	43	44	45	46	47	47	48	49	49
50	1	2	2	3	4	5	6	7	8	8	9	10	11	12	12	13	14	15	16	17	18	18	19	20	21	22	22	23	24	25	26	27	28	28	29	30	31	32	32	33	34	35	36	37	38	38	39	40	41	42	42	43	44	45	46	47	48	48	49	50	50
51	1	2	3	3	4	5	6	7	8	8	9	10	11	12	13	14	14	15	16	17	18	19	20	20	21	22	23	24	25	26	26	27	28	29	30	31	31	32	33	34	35	36	37	37	38	39	40	41	42	42	43	44	45	46	47	48	48	49	50	51	51
52	1	2	3	3	4	5	6	7	8	9	10	10	11	12	13	14	15	16	16	17	18	19	20	21	22	23	23	24	25	26	27	28	29	29	30	31	32	33	34	35	36	36	37	38	39	40	41	42	42	43	44	45	46	47	48	49	49	50	51	52	52
53	1	2	3	4	4	5	6	7	8	9	10	11	11	12	13	14	15	16	17	18	19	19	20	21	22	23	24	25	26	26	27	28	29	30	31	32	33	34	34	35	36	37	38	39	40	41	42	42	43	44	45	46	47	48	49	49	50	51	52	53	53
54	1	2	3	4	4	5	6	7	8	9	10	11	12	13	14	14	15	16	17	18	19	20	21	22	22	23	24	25	26	27	28	29	30	31	32	32	33	34	35	36	37	38	39	40	40	41	42	43	44	45	46	47	48	49	50	50	51	52	53	54	54
55	1	2	3	4	5	6	6	7	8	9	10	11	12	13	14	15	16	16	17	18	19	20	21	22	23	24	25	26	27	28	28	29	30	31	32	33	34	35	36	37	38	38	39	40	41	42	43	44	45	46	47	48	49	50	50	51	52	53	54	55	55
56	1	2	3	4	5	6	7	7	8	9	10	11	12	13	14	15	16	17	18	19	20	21	21	22	23	24	25	26	27	28	29	30	31	32	33	34	35	35	36	37	38	39	40	41	42	43	44	45	46	47	48	49	49	50	51	52	53	54	55	56	56
57	1	2	3	4	5	6	7	8	9	10	10	11	12	13	14	15	16	17	18	19	20	21	22	23	24	25	26	27	28	28	29	30	31	32	33	34	35	36	37	38	39	40	41	42	43	44	45	46	47	48	48	49	50	51	52	53	54	55	56	57	57
58	1	2	3	4	5	6	7	8	9	10	11	12	13	14	14	15	16	17	18	19	20	21	22	23	24	25	26	27	28	29	30	31	32	33	34	35	36	37	38	39	40	41	42	43	44	44	45	46	47	48	49	50	51	52	53	54	55	56	57	58	58
59	1	2	3	4	5	6	7	8	9	10	11	12	13	14	15	16	17	18	19	20	21	22	23	24	25	26	27	28	29	30	30	31	32	33	34	35	36	37	38	39	40	41	42	43	44	45	46	47	48	49	50	51	52	53	54	55	56	57	58	59	59

LAT 41°

N. Lat. {LHA greater than 180°........ Zn=Z; LHA less than 180°............ Zn=360−Z}

DECLINATION (0°-14°) SAME NAME AS LATITUDE

LHA	0° Hc	d	Z	1° Hc	d	Z	2° Hc	d	Z	3° Hc	d	Z	4° Hc	d	Z
	° ′	′	°	° ′	′	°	° ′	′	°	° ′	′	°	° ′	′	°
0	49 00	+60	180	50 00	+60	180	51 00	+60	180	52 00	+60	180	53 00	+60	180
1	48 59	60	179	49 59	60	178	50 59	60	178	51 59	60	178	52 59	60	178
2	48 58	60	177	49 58	60	177	50 58	59	177	51 57	60	177	52 57	60	177
3	48 55	60	175	49 55	59	175	50 54	60	175	51 54	60	175	52 54	60	175
4	48 50	60	174	49 50	60	174	50 50	60	174	51 50	60	174	52 50	59	173
5	48 45	+60	172	49 45	+59	172	50 44	+60	172	51 44	+60	172	52 44	+59	172
6	48 38	60	171	49 38	60	171	50 38	59	171	51 37	60	170	52 37	59	170
7	48 31	59	169	49 30	60	169	50 30	59	169	51 29	59	169	52 28	59	169
8	48 22	59	168	49 21	59	168	50 20	59	167	51 19	60	167	52 19	59	167
9	48 12	59	166	49 11	59	166	50 10	59	166	51 09	59	166	52 08	58	165
10	48 01	+58	165	48 59	+59	165	49 58	+59	164	50 57	+59	164	51 56	+58	164
11	47 48	59	164	48 47	58	163	49 45	59	163	50 44	58	163	51 42	59	162
12	47 35	58	162	48 33	58	162	49 31	59	161	50 30	58	161	51 28	58	161
13	47 20	58	161	48 18	58	160	49 16	58	160	50 14	58	159	51 12	58	159
14	47 05	58	159	48 03	57	159	49 00	58	158	49 58	57	158	50 55	58	158
15	46 48	+58	158	47 46	+57	157	48 43	+57	157	49 40	+57	157	50 37	+57	156
16	46 31	57	156	47 28	57	156	48 25	57	156	49 22	56	155	50 18	57	155
17	46 12	57	155	47 09	56	155	48 05	57	154	49 02	56	154	49 58	57	153
18	45 52	57	154	46 49	56	153	47 45	56	153	48 41	56	152	49 37	56	152
19	45 32	56	152	46 28	56	152	47 24	56	151	48 20	55	151	49 15	56	150
20	45 10	+56	151	46 06	+56	151	47 02	+55	150	47 57	+55	149	48 52	+55	149
21	44 48	55	150	45 43	55	149	46 38	55	149	47 33	55	148	48 28	55	147
22	44 24	56	148	45 20	54	148	46 14	55	147	47 09	54	147	48 03	55	146
23	44 00	55	147	44 55	54	147	45 49	55	146	46 44	54	145	47 38	54	145
24	43 35	55	146	44 30	54	145	45 24	53	145	46 17	54	144	47 11	54	143
25	43 09	+54	145	44 03	+54	144	44 57	+53	143	45 50	+54	143	46 44	+53	142
26	42 43	53	143	43 36	54	143	44 30	53	142	45 23	52	141	46 15	53	141
27	42 15	54	142	43 09	52	142	44 01	53	141	44 54	52	140	45 46	53	140
28	41 47	53	141	42 40	52	140	43 32	53	140	44 25	52	139	45 17	51	138
29	41 18	53	140	42 11	52	139	43 03	52	139	43 55	51	138	44 46	51	137
30	40 49	+52	139	41 41	+51	138	42 32	+52	137	43 24	+51	137	44 15	+51	136
31	40 19	51	138	41 10	51	137	42 01	51	136	42 52	51	135	43 43	51	135
32	39 48	51	136	40 39	51	136	41 30	50	135	42 20	51	134	43 11	50	134
33	39 16	51	135	40 07	50	135	40 57	51	134	41 48	49	133	42 37	50	132
34	38 44	50	134	39 34	50	134	40 24	50	133	41 14	50	132	42 04	49	131
35	38 11	+50	133	39 01	+50	132	39 51	+49	132	40 40	+49	131	41 29	+49	130
36	37 38	49	132	38 27	50	131	39 17	49	131	40 06	49	130	40 55	48	129
37	37 04	49	131	37 53	49	130	38 42	49	130	39 31	48	129	40 19	48	128
38	36 30	48	130	37 18	49	129	38 07	48	129	38 55	48	128	39 43	48	127
39	35 55	48	129	36 43	48	128	37 31	48	128	38 19	48	127	39 07	47	126
40	35 19	+48	128	36 07	+48	127	36 55	+48	127	37 43	+47	126	38 30	+47	125
41	34 43	48	127	35 31	48	126	36 19	47	126	37 06	47	125	37 53	46	124
42	34 07	47	126	34 54	48	125	35 42	46	125	36 28	47	124	37 15	46	123
43	33 30	47	125	34 17	47	124	35 04	47	124	35 51	46	123	36 37	46	122
44	32 53	47	124	33 40	46	123	34 26	46	123	35 12	46	122	35 58	46	121
45	32 15	+47	123	33 02	+46	123	33 48	+46	122	34 34	+45	121	35 19	+45	120
46	31 37	46	122	32 23	46	122	33 09	46	121	33 55	45	120	34 40	45	119
47	30 59	46	122	31 45	45	121	32 30	45	120	33 15	45	119	34 00	45	118
48	30 20	45	121	31 05	46	120	31 51	45	119	32 36	44	118	33 20	44	118
49	29 41	45	120	30 26	45	119	31 11	44	118	31 55	45	117	32 40	44	117
50	29 01	+45	119	29 46	+45	118	30 31	+44	117	31 15	+44	117	31 59	+44	116
51	28 21	45	118	29 06	44	117	29 50	44	116	30 34	44	116	31 18	44	115
52	27 41	45	117	28 26	44	116	29 10	43	116	29 53	44	115	30 37	43	114
53	27 01	44	116	27 45	44	116	28 29	43	115	29 12	43	114	29 55	43	113
54	26 20	44	116	27 04	43	115	27 47	44	114	28 31	43	113	29 14	42	112
55	25 39	+44	115	26 23	+43	114	27 06	+43	113	27 49	+43	112	28 32	+42	112
56	24 58	43	114	25 41	43	113	26 24	43	112	27 07	43	112	27 50	42	111
57	24 16	43	113	24 59	43	112	25 42	43	112	26 25	42	111	27 07	42	110
58	23 34	43	112	24 17	43	112	25 00	42	111	25 42	42	110	26 24	42	109
59	22 52	43	112	23 35	42	111	24 17	43	110	25 00	42	109	25 42	41	108
60	22 10	+43	111	22 53	+42	110	23 35	+42	109	24 17	+41	108	24 58	+42	108
61	21 28	42	110	22 10	42	109	22 52	42	108	23 34	41	108	24 15	41	107
62	20 45	42	109	21 27	42	109	22 09	41	108	22 50	42	107	23 32	41	106
63	20 02	42	109	20 44	42	108	21 26	41	107	22 07	41	106	22 48	41	105
64	19 19	42	108	20 01	41	107	20 42	41	106	21 23	41	105	22 04	41	105
65	18 36	+41	107	19 17	+42	106	19 59	+41	106	20 40	+41	105	21 21	+40	104
66	17 53	41	106	18 34	41	106	19 15	41	105	19 56	41	104	20 37	40	103
67	17 09	41	106	17 50	41	105	18 31	41	104	19 12	40	103	19 52	41	103
68	16 25	41	105	17 06	41	104	17 47	41	103	18 28	40	103	19 08	40	102
69	15 42	40	104	16 22	41	103	17 03	40	103	17 43	41	102	18 24	40	101

LHA	5° Hc	d	Z	6° Hc	d	Z	7° Hc	d	Z	8° Hc	d	Z	9° Hc	d	Z
	° ′	′	°	° ′	′	°	° ′	′	°	° ′	′	°	° ′	′	°
0	54 00	+60	180	55 00	+60	180	56 00	+60	180	57 00	+60	180	58 00	+60	180
1	53 59	60	178	54 59	60	178	55 59	60	178	56 59	60	178	57 59	60	178
2	53 57	60	177	54 57	60	177	55 57	60	177	56 57	60	176	57 57	60	176
3	53 54	60	175	54 54	60	175	55 54	60	175	56 54	59	175	57 53	60	174
4	53 49	60	173	54 49	60	173	55 49	60	173	56 49	59	173	57 48	60	173
5	53 43	+60	172	54 43	+60	171	55 43	+59	171	56 42	+60	171	57 42	+59	171
6	53 36	60	170	54 36	59	170	55 35	59	169	56 34	60	169	57 34	59	169
7	53 27	60	168	54 27	59	168	55 26	59	168	56 25	59	167	57 24	59	167
8	53 18	59	167	54 17	59	166	55 16	59	166	56 15	58	166	57 13	59	165
9	53 06	59	165	54 05	59	165	55 04	59	164	56 03	58	164	57 01	59	164
10	52 54	+59	163	53 53	+58	163	54 51	+58	163	55 49	+59	162	56 48	+58	162
11	52 41	58	162	53 39	58	161	54 37	58	161	55 35	58	161	56 33	58	160
12	52 26	58	160	53 24	57	160	54 21	58	159	55 19	58	159	56 17	57	158
13	52 10	57	159	53 07	58	158	54 05	57	158	55 02	57	157	55 59	58	157
14	51 53	57	157	52 50	57	157	53 47	57	156	54 44	57	156	55 41	57	155
15	51 34	+57	156	52 31	+57	155	53 28	+57	154	54 25	+56	154	55 21	+56	153
16	51 15	57	154	52 12	56	153	53 08	56	153	54 04	56	152	55 00	56	152
17	50 55	56	153	51 51	56	152	52 47	56	151	53 43	55	151	54 38	55	150
18	50 33	56	151	51 29	56	150	52 25	55	150	53 20	55	149	54 15	55	149
19	50 11	55	150	51 06	55	149	52 01	55	148	52 56	55	148	53 51	54	147
20	49 47	+55	148	50 42	+55	148	51 37	+54	147	52 31	+55	146	53 26	+54	146
21	49 23	55	147	50 18	54	146	51 12	54	145	52 06	53	145	52 59	54	144
22	48 58	54	145	49 52	54	145	50 46	53	144	51 39	53	143	52 32	53	143
23	48 32	53	144	49 25	54	143	50 19	53	143	51 12	52	142	52 04	53	141
24	48 05	53	143	48 58	53	142	49 51	52	141	50 43	53	141	51 36	52	140
25	47 37	+52	141	48 29	+53	141	49 22	+52	140	50 14	+52	139	51 06	+51	138
26	47 08	52	140	48 00	52	139	48 52	52	139	49 44	51	138	50 35	51	137
27	46 39	51	139	47 30	52	138	48 22	51	137	49 13	51	137	50 04	51	136
28	46 08	52	138	47 00	51	137	47 51	51	136	48 42	50	135	49 32	50	134
29	45 37	51	136	46 28	51	136	47 19	50	135	48 09	50	134	48 59	50	133
30	45 06	+50	135	45 56	+51	134	46 47	+49	134	47 36	+50	133	48 26	+49	132
31	44 34	50	134	45 24	49	133	46 13	50	132	47 03	49	132	47 52	49	131
32	44 01	49	133	44 50	50	132	45 40	49	131	46 29	48	130	47 17	49	130
33	43 27	49	132	44 16	49	131	45 05	49	130	45 54	48	129	46 42	48	128
34	42 53	49	131	43 42	48	130	44 30	49	129	45 19	47	128	46 06	48	127
35	42 18	+49	129	43 07	+48	129	43 55	+48	128	44 43	+47	127	45 30	+47	126
36	41 43	48	128	42 31	48	128	43 19	47	127	44 06	47	126	44 53	47	125
37	41 07	48	127	41 55	47	127	42 42	47	126	43 29	47	125	44 16	46	124
38	40 31	47	126	41 18	47	125	42 05	47	125	42 52	46	124	43 38	46	123
39	39 54	47	125	40 41	47	124	41 28	46	124	42 14	46	123	43 00	45	122
40	39 17	+46	124	40 03	+47	123	40 50	+46	123	41 36	+45	122	42 21	+45	121
41	38 39	46	123	39 25	46	122	40 11	46	122	40 57	45	121	41 42	45	120
42	38 01	46	122	38 47	46	121	39 33	45	121	40 18	44	120	41 02	45	119
43	37 23	45	121	38 08	45	120	38 53	45	120	39 38	45	119	40 23	44	118
44	36 44	45	120	37 29	45	120	38 14	44	119	38 58	44	118	39 42	44	117
45	36 04	+45	119	36 49	+45	119	37 34	+44	118	38 18	+44	117	39 02	+43	116
46	35 25	44	118	36 09	45	118	36 54	43	117	37 37	44	116	38 21	43	115
47	34 45	44	118	35 29	44	117	36 13	44	116	36 57	43	115	37 40	43	114
48	34 04	44	117	34 48	44	116	35 32	43	115	36 15	43	114	36 58	43	113
49	33 24	44	116	34 08	43	115	34 51	43	114	35 34	43	113	36 17	42	112
50	32 43	+43	115	33 26	+43	114	34 09	+43	113	34 52	+43	112	35 35	+42	112
51	32 02	43	114	32 45	43	113	33 28	42	112	34 10	42	112	34 52	42	111
52	31 20	43	113	32 03	43	112	32 46	42	112	33 28	42	111	34 10	41	110
53	30 38	43	112	31 21	42	112	32 03	42	111	32 45	42	110	33 27	41	109
54	29 56	43	112	30 39	42	111	31 21	42	110	32 03	41	109	32 44	41	108
55	29 14	+42	111	29 56	+42	110	30 38	+42	109	31 20	+41	108	32 01	+41	107
56	28 32	42	110	29 14	41	109	29 55	42	108	30 37	41	108	31 18	41	107
57	27 49	42	109	28 31	41	108	29 12	42	108	29 54	40	107	30 34	41	106
58	27 06	42	108	27 48	41	108	28 29	41	107	29 10	41	106	29 51	40	105
59	26 23	42	108	27 05	41	107	27 46	40	106	28 26	41	105	29 07	40	104
60	25 40	+41	107	26 21	+41	106	27 02	+41	105	27 43	+40	104	28 23	+40	104
61	24 56	41	106	25 37	41	105	26 18	41	105	26 59	40	104	27 39	40	103
62	24 13	41	105	24 54	40	105	25 34	41	104	26 15	40	103	26 55	39	102
63	23 29	41	105	24 10	40	104	24 50	40	103	25 30	40	102	26 10	40	101
64	22 45	41	104	23 26	40	103	24 06	40	102	24 46	40	101	25 26	39	101
65	22 01	+41	103	22 42	+40	102	23 22	+40	102	24 02	+39	101	24 41	+40	100
66	21 17	40	102	21 57	40	102	22 37	40	101	23 17	40	100	23 57	39	99
67	20 33	40	102	21 13	40	101	21 53	39	100	22 32	40	99	23 12	39	99
68	19 48	40	101	20 28	40	100	21 08	40	99	21 48	39	99	22 27	39	98
69	19 04	40	100	19 44	39	100	20 23	40	99	21 03	39	98	21 42	39	97

LHA	10° Hc	d	Z	11° Hc	d	Z	12° Hc	d	Z	13° Hc	d	Z	14° Hc	d	Z	LHA
	° ′	′	°	° ′	′	°	° ′	′	°	° ′	′	°	° ′	′	°	
0	59 00	+60	180	60 00	+60	180	61 00	+60	180	62 00	+60	180	63 00	+60	180	360
1	58 59	60	178	59 59	60	178	60 59	60	178	61 59	60	178	62 59	60	178	359
2	58 57	60	176	59 57	60	176	60 57	60	176	61 57	60	176	62 57	60	176	358
3	58 53	60	174	59 53	60	174	60 53	60	174	61 53	59	174	62 52	60	174	357
4	58 48	60	172	59 48	59	172	60 47	60	172	61 47	60	172	62 47	59	172	356
5	58 41	+60	171	59 41	+59	170	60 40	+60	170	61 40	+59	170	62 39	+59	169	355
6	58 33	59	169	59 32	60	168	60 32	59	168	61 31	59	168	62 30	59	167	354
7	58 23	59	167	59 22	59	166	60 21	59	166	61 20	59	166	62 19	59	165	353
8	58 12	59	165	59 11	59	165	60 10	58	164	61 08	59	164	62 07	58	163	352
9	58 00	59	163	58 58	59	163	59 57	58	162	60 55	58	162	61 53	58	161	351
10	57 46	+58	161	58 44	+58	161	59 42	+58	160	60 40	+58	160	61 38	+57	159	350
11	57 31	58	160	58 29	57	159	59 26	58	159	60 24	57	158	61 21	57	157	349
12	57 14	58	158	58 12	57	157	59 09	57	157	60 06	57	156	61 03	56	155	348
13	56 57	57	156	57 54	56	156	58 50	57	155	59 47	56	154	60 43	56	154	347
14	56 38	56	154	57 34	56	154	58 30	56	153	59 26	56	152	60 22	56	152	346
15	56 17	+56	153	57 13	+56	152	58 09	+56	151	59 05	+55	151	60 00	+55	150	345
16	55 56	56	151	56 52	55	150	57 47	55	150	58 42	55	149	59 37	54	148	344
17	55 33	56	149	56 29	54	149	57 23	55	148	58 18	54	147	59 12	54	146	343
18	55 10	55	148	56 05	54	147	56 59	54	146	57 53	54	146	58 47	53	145	342
19	54 45	54	146	55 39	54	146	56 33	54	145	57 27	53	144	58 20	53	143	341
20	54 20	+53	145	55 13	+54	144	56 07	+53	143	57 00	+52	142	57 52	+53	141	340
21	53 53	53	143	54 46	53	142	55 39	52	142	56 31	53	141	57 24	51	140	339
22	53 25	53	142	54 18	52	141	55 10	52	140	56 02	52	139	56 54	51	138	338
23	52 57	52	140	53 49	52	140	54 41	51	139	55 32	51	138	56 23	51	137	337
24	52 28	51	139	53 19	51	138	54 10	51	137	55 01	51	136	55 52	50	135	336
25	51 57	+52	138	52 49	+50	137	53 39	+51	136	54 30	+50	135	55 20	+49	134	335
26	51 26	51	136	52 17	50	135	53 07	50	134	53 57	50	134	54 47	48	133	334
27	50 55	50	135	51 45	50	134	52 35	49	133	53 24	49	132	54 13	48	131	333
28	50 22	50	134	51 12	49	133	52 01	49	132	52 50	48	131	53 38	48	130	332
29	49 49	49	132	50 38	49	131	51 27	48	131	52 15	48	130	53 03	48	129	331
30	49 15	+49	131	50 04	+48	130	50 52	+48	129	51 40	+47	128	52 27	+47	127	330
31	48 41	48	130	49 29	48	129	50 17	47	128	51 04	47	127	51 51	46	126	329
32	48 06	47	129	48 53	48	128	49 41	47	127	50 28	46	126	51 14	46	125	328
33	47 30	47	127	48 17	47	127	49 04	47	126	49 51	46	125	50 37	45	124	327
34	46 54	47	126	47 41	46	125	48 27	46	124	49 13	46	124	49 59	45	123	326
35	46 17	+47	125	47 04	+46	124	47 50	+45	123	48 35	+45	122	49 20	+45	121	325
36	45 40	46	124	46 26	46	123	47 12	45	122	47 57	44	121	48 41	45	120	324
37	45 02	46	123	45 48	45	122	46 33	45	121	47 18	44	120	48 02	44	119	323
38	44 24	45	122	45 09	45	121	45 54	44	120	46 38	44	119	47 22	44	118	322
39	43 45	45	121	44 30	45	120	45 15	44	119	45 59	43	118	46 42	43	117	321
40	43 06	+45	120	43 51	+44	119	44 35	+44	118	45 19	+43	117	46 02	+42	116	320
41	42 27	44	119	43 11	44	118	43 55	43	117	44 38	43	116	45 21	42	115	319
42	41 47	44	118	42 31	43	117	43 14	43	116	43 57	43	115	44 40	42	114	318
43	41 07	43	117	41 50	43	116	42 33	43	115	43 16	42	114	43 58	42	113	317
44	40 26	43	116	41 09	43	115	41 52	43	114	42 35	41	113	43 16	42	112	316
45	39 45	+43	115	40 28	+43	114	41 11	+42	113	41 53	+41	112	42 34	+41	111	315
46	39 04	43	114	39 47	42	113	40 29	42	112	41 11	41	111	41 52	41	110	314
47	38 23	42	113	39 05	42	112	39 47	41	111	40 28	42	111	41 10	40	110	313
48	37 41	42	112	38 23	42	112	39 05	41	111	39 46	41	110	40 27	40	109	312
49	36 59	42	112	37 41	41	111	38 22	41	110	39 03	41	109	39 44	40	108	311
50	36 17	+41	111	36 58	+41	110	37 39	+41	109	38 20	+41	108	39 01	+39	107	310
51	35 34	41	110	36 15	41	109	36 56	41	108	37 37	40	107	38 17	40	106	309
52	34 51	42	109	35 33	40	108	36 13	41	107	36 54	40	106	37 34	39	105	308
53	34 08	41	108	34 49	41	107	35 30	40	106	36 10	40	105	36 50	39	105	307
54	33 25	41	107	34 06	40	107	34 46	40	106	35 26	40	105	36 06	39	104	306
55	32 42	+41	107	33 23	+40	106	34 03	+39	105	34 42	+40	104	35 22	+39	103	305
56	31 59	40	106	32 39	40	105	33 19	39	104	33 58	40	103	34 38	38	102	304
57	31 15	40	105	31 55	40	104	32 35	39	103	33 14	39	102	33 53	39	101	303
58	30 31	40	104	31 11	40	103	31 51	39	103	32 30	39	102	33 09	38	101	302
59	29 47	40	103	30 27	39	103	31 06	39	102	31 45	39	101	32 24	39	100	301
60	29 03	+40	103	29 43	+39	102	30 22	+39	101	31 01	+39	100	31 40	+38	99	300
61	28 19	39	102	28 58	39	101	29 37	39	100	30 16	39	99	30 55	38	98	299
62	27 34	40	101	28 14	39	100	28 53	39	100	29 32	38	99	30 10	38	98	298
63	26 50	39	101	27 29	39	100	28 08	39	99	28 47	38	98	29 25	38	97	297
64	26 05	39	100	26 44	39	99	27 23	39	98	28 02	38	97	28 40	38	96	296
65	25 21	+39	99	26 00	+38	98	26 38	+39	97	27 17	+38	97	27 55	+38	96	295
66	24 36	39	98	25 15	39	98	25 54	38	97	26 32	38	96	27 10	38	95	294
67	23 51	39	98	24 30	39	97	25 09	38	96	25 47	38	95	26 25	38	94	293
68	23 06	39	97	23 45	38	96	24 23	39	95	25 02	38	94	25 40	37	94	292
69	22 21	39	96	23 00	38	95	23 38	39	95	24 17	37	94	24 54	38	93	291

S. Lat. {LHA greater than 180°........ Zn=180−Z; LHA less than 180°............ Zn=180+Z}

DECLINATION (0°-14°) SAME NAME AS LATITUDE

LAT 41°

LAT 41°

DECLINATION (15°-29°) SAME NAME AS LATITUDE

N. Lat. {LHA greater than 180°....... Zn=Z; LHA less than 180°........... Zn=360−Z}

LHA	15° Hc	d	Z	16° Hc	d	Z	17° Hc	d	Z	18° Hc	d	Z	19° Hc	d	Z
0	64 00	+60	180	65 00	+60	180	66 00	+60	180	67 00	+60	180	68 00	+60	180
1	63 59	60	178	64 59	60	178	65 59	60	178	66 59	60	178	67 59	60	178
2	63 57	59	176	64 56	60	176	65 56	60	175	66 56	60	175	67 56	60	175
3	63 52	60	173	64 52	60	173	65 52	59	173	66 51	60	173	67 51	60	173
4	63 46	60	171	64 46	59	171	65 45	60	171	66 45	59	170	67 44	60	170
5	63 38	+60	169	64 38	+59	169	65 37	+59	168	66 36	+59	168	67 35	+59	168
6	63 29	59	167	64 28	59	167	65 27	59	166	66 26	59	166	67 25	58	165
7	63 18	59	165	64 17	58	164	65 15	59	164	66 14	58	163	67 12	58	163
8	63 05	59	163	64 04	58	162	65 02	58	162	66 00	59	161	66 58	57	160
9	62 51	58	161	63 49	58	160	64 47	57	159	65 44	58	159	66 42	57	158
10	62 35	+58	159	63 33	+57	158	64 30	+57	157	65 27	+57	157	66 24	+56	156
11	62 18	57	157	63 15	57	156	64 12	56	155	65 08	57	154	66 05	55	154
12	61 59	57	155	62 56	56	154	63 52	56	153	64 48	56	152	65 44	55	151
13	61 39	56	153	62 35	56	152	63 31	55	151	64 26	55	150	65 21	55	149
14	61 18	55	151	62 13	55	150	63 08	55	149	64 03	54	148	64 57	54	147
15	60 55	+55	149	61 50	+55	148	62 45	+54	147	63 39	+53	146	64 32	+54	145
16	60 31	55	147	61 26	53	146	62 19	54	145	63 13	53	144	64 06	52	143
17	60 06	54	146	61 00	53	145	61 53	53	144	62 46	52	143	63 38	52	142
18	59 40	53	144	60 33	53	143	61 26	52	142	62 18	52	141	63 10	51	140
19	59 13	52	142	60 05	52	141	60 57	52	140	61 49	51	139	62 40	50	138
20	58 45	+51	141	59 36	+52	140	60 28	+51	138	61 19	+50	137	62 09	+50	136
21	58 15	51	139	59 06	51	138	59 57	50	137	60 47	50	136	61 37	49	135
22	57 45	51	137	58 36	50	136	59 26	49	135	60 15	49	134	61 04	49	133
23	57 14	50	136	58 04	49	135	58 53	49	134	59 42	49	133	60 31	47	131
24	56 42	49	134	57 31	49	133	58 20	49	132	59 09	47	131	59 56	47	130
25	56 09	+49	133	56 58	+48	132	57 46	+48	131	58 34	+47	130	59 21	+47	128
26	55 35	49	132	56 24	48	130	57 12	47	129	57 59	46	128	58 45	46	127
27	55 01	48	130	55 49	47	129	56 36	47	128	57 23	46	127	58 09	45	126
28	54 26	47	129	55 13	47	128	56 00	46	127	56 46	46	125	57 32	44	124
29	53 51	46	128	54 37	47	126	55 24	45	125	56 09	45	124	56 54	44	123
30	53 14	+47	126	54 01	+45	125	54 46	+45	124	55 31	+45	123	56 16	+43	122
31	52 37	46	125	53 23	45	124	54 08	45	123	54 53	44	122	55 37	43	120
32	52 00	45	124	52 45	45	123	53 30	44	122	54 14	44	120	54 58	42	119
33	51 22	45	123	52 07	44	122	52 51	44	120	53 35	43	119	54 18	42	118
34	50 44	44	121	51 28	44	120	52 12	43	119	52 55	43	118	53 38	42	117
35	50 05	+44	120	50 49	+43	119	51 32	+43	118	52 15	+42	117	52 57	+42	116
36	49 26	43	119	50 09	43	118	50 52	43	117	51 35	41	116	52 16	41	115
37	48 46	43	118	49 29	43	117	50 12	42	116	50 54	41	115	51 35	41	114
38	48 06	43	117	48 49	42	116	49 31	41	115	50 12	41	114	50 53	41	113
39	47 25	43	116	48 08	42	115	48 50	41	114	49 31	40	113	50 11	40	112
40	46 44	+42	115	47 26	+42	114	48 08	+41	113	48 49	+40	112	49 29	+40	111
41	46 03	42	114	46 45	41	113	47 26	41	112	48 07	40	111	48 47	39	110
42	45 22	41	113	46 03	41	112	46 44	40	111	47 24	40	110	48 04	39	109
43	44 40	41	112	45 21	41	111	46 02	40	110	46 42	39	109	47 21	39	108
44	43 58	41	111	44 39	40	110	45 19	40	109	45 59	39	108	46 38	38	107
45	43 15	+41	110	43 56	+40	109	44 36	+39	108	45 15	+39	107	45 54	+38	106
46	42 33	40	109	43 13	40	108	43 53	39	107	44 32	39	106	45 11	38	105
47	41 50	40	109	42 30	40	108	43 10	39	107	43 49	38	106	44 27	38	104
48	41 07	40	108	41 47	39	107	42 26	39	106	43 05	38	105	43 43	38	104
49	40 24	39	107	41 03	39	106	41 42	39	105	42 21	38	104	42 59	37	103
50	39 40	+40	106	40 20	+39	105	40 59	+38	104	41 37	+38	103	42 15	+37	102
51	38 57	39	105	39 36	39	104	40 15	38	103	40 53	37	102	41 30	37	101
52	38 13	39	104	38 52	38	103	39 30	38	102	40 08	38	101	40 46	37	100
53	37 29	39	104	38 08	38	103	38 46	38	102	39 24	37	101	40 01	37	100
54	36 45	39	103	37 24	38	102	38 02	37	101	38 39	37	100	39 16	37	99
55	36 01	+38	102	36 39	+38	101	37 17	+38	100	37 55	+37	99	38 32	+36	98
56	35 16	39	101	35 55	38	100	36 33	37	99	37 10	37	98	37 47	36	97
57	34 32	38	101	35 10	38	100	35 48	37	99	36 25	37	98	37 02	36	97
58	33 47	38	100	34 25	38	99	35 03	37	98	35 40	37	97	36 17	36	96
59	33 03	38	99	33 41	37	98	34 18	37	97	34 55	37	96	35 32	36	95
60	32 18	+38	98	32 56	+37	97	33 33	+37	96	34 10	+37	96	34 47	+36	95
61	31 33	38	98	32 11	37	97	32 48	37	96	33 25	37	95	34 02	36	94
62	30 48	38	97	31 26	37	96	32 03	37	95	32 40	36	94	33 16	36	93
63	30 03	38	96	30 41	37	95	31 18	37	94	31 55	36	93	32 31	36	92
64	29 18	38	95	29 56	37	95	30 33	36	94	31 09	37	93	31 46	36	92
65	28 33	+37	95	29 10	+37	94	29 47	+37	93	30 24	+37	92	31 01	+36	91
66	27 48	37	94	28 25	37	93	29 02	37	92	29 39	36	91	30 15	36	90
67	27 03	37	93	27 40	37	93	28 17	37	92	28 54	36	91	29 30	36	90
68	26 17	38	93	26 55	37	92	27 32	36	91	28 08	37	90	28 45	36	89
69	25 32	37	92	26 09	37	91	26 46	37	90	27 23	36	89	27 59	36	89

LHA	20° Hc	d	Z	21° Hc	d	Z	22° Hc	d	Z	23° Hc	d	Z	24° Hc	d	Z
0	69 00	+60	180	70 00	+60	180	71 00	+60	180	72 00	+60	180	73 00	+60	180
1	68 59	60	177	69 59	60	177	70 59	60	177	71 59	60	177	72 59	60	177
2	68 56	60	175	69 56	60	175	70 56	59	174	71 55	60	174	72 55	60	174
3	68 51	59	172	69 50	60	172	70 50	60	172	71 50	59	171	72 49	59	171
4	68 44	59	170	69 43	59	169	70 42	59	169	71 41	59	168	72 40	59	168
5	68 34	+59	167	69 33	+59	167	70 32	+59	166	71 31	+59	165	72 30	+58	165
6	68 23	59	165	69 22	58	164	70 20	58	163	71 18	59	163	72 17	57	162
7	68 10	58	162	69 08	58	161	70 06	58	161	71 04	57	160	72 01	57	159
8	67 55	58	160	68 53	57	159	69 50	57	158	70 47	57	157	71 44	56	156
9	67 39	57	157	68 36	56	156	69 32	57	156	70 29	56	155	71 25	55	153
10	67 20	+57	155	68 17	+56	154	69 13	+55	153	70 08	+55	152	71 03	+55	151
11	67 00	56	153	67 56	55	152	68 51	55	151	69 46	54	150	70 40	54	148
12	66 39	55	151	67 34	54	149	68 28	54	148	69 22	54	147	70 16	52	146
13	66 16	54	148	67 10	54	147	68 04	53	146	68 57	52	145	69 49	52	143
14	65 51	54	146	66 45	53	145	67 38	52	144	68 30	52	143	69 22	51	141
15	65 26	+52	144	66 18	+52	143	67 10	+52	142	68 02	+51	140	68 53	+50	139
16	64 58	52	142	65 50	52	141	66 42	50	140	67 32	50	138	68 22	49	137
17	64 30	51	140	65 21	51	139	66 12	50	138	67 02	49	136	67 51	48	135
18	64 01	50	139	64 51	50	137	65 41	49	136	66 30	48	135	67 18	48	133
19	63 30	50	137	64 20	49	135	65 09	48	134	65 57	48	133	66 45	46	131
20	62 59	+49	135	63 48	+48	134	64 36	+47	132	65 23	+47	131	66 10	+46	129
21	62 26	48	133	63 14	48	132	64 02	47	131	64 49	46	129	65 35	44	128
22	61 53	47	132	62 40	47	130	63 27	46	129	64 13	45	128	64 58	44	126
23	61 18	47	130	62 05	47	129	62 52	45	127	63 37	44	126	64 21	44	124
24	60 43	47	129	61 30	45	127	62 15	45	126	63 00	44	124	63 44	42	123
25	60 08	+45	127	60 53	+45	126	61 38	+44	124	62 22	+43	123	63 05	+42	122
26	59 31	45	126	60 16	45	124	61 01	43	123	61 44	42	122	62 26	42	120
27	58 54	45	124	59 39	43	123	60 22	43	122	61 05	42	120	61 47	41	119
28	58 16	44	123	59 00	44	122	59 44	42	120	60 26	41	119	61 07	40	117
29	57 38	44	122	58 22	42	120	59 04	42	119	59 46	41	118	60 27	39	116
30	56 59	+43	120	57 42	+42	119	58 24	+41	118	59 05	+41	116	59 46	+39	115
31	56 20	42	119	57 02	42	118	57 44	41	117	58 25	39	115	59 04	39	114
32	55 40	42	118	56 22	41	117	57 03	41	115	57 44	39	114	58 23	38	113
33	55 00	42	117	55 42	40	116	56 22	40	114	57 02	39	113	57 41	38	112
34	54 20	41	116	55 01	40	114	55 41	39	113	56 20	39	112	56 59	37	110
35	53 39	+40	115	54 19	+40	113	54 59	+39	112	55 38	+38	111	56 16	+37	109
36	52 57	40	114	53 37	40	112	54 17	38	111	54 55	38	110	55 33	37	108
37	52 16	39	113	52 55	39	111	53 34	39	110	54 13	37	109	54 50	36	107
38	51 34	39	112	52 13	39	110	52 52	38	109	53 30	37	108	54 07	36	106
39	50 51	39	111	51 30	39	109	52 09	37	108	52 46	37	107	53 23	36	105
40	50 09	+39	110	50 48	+38	108	51 26	+37	107	52 03	+36	106	52 39	+36	105
41	49 26	38	109	50 04	38	107	50 42	37	106	51 19	36	105	51 55	36	104
42	48 43	38	108	49 21	38	107	49 59	36	105	50 35	36	104	51 11	35	103
43	48 00	38	107	48 38	37	106	49 15	36	104	49 51	36	103	50 27	35	102
44	47 16	38	106	47 54	37	105	48 31	36	104	49 07	36	102	49 43	34	101
45	46 32	+38	105	47 10	+37	104	47 47	+36	103	48 23	+35	102	48 58	+35	100
46	45 49	37	104	46 26	37	103	47 03	35	102	47 38	36	101	48 14	34	99
47	45 05	37	103	45 42	36	102	46 18	36	101	46 54	35	100	47 29	34	99
48	44 21	36	103	44 57	37	101	45 34	35	100	46 09	35	99	46 44	34	98
49	43 36	37	102	44 13	36	101	44 49	35	99	45 24	35	98	45 59	34	97
50	42 52	+36	101	43 28	+36	100	44 04	+36	99	44 40	+34	98	45 14	+34	96
51	42 07	37	100	42 44	36	99	43 20	35	98	43 55	34	97	44 29	34	96
52	41 23	36	99	41 59	36	98	42 35	35	97	43 10	34	96	43 44	34	95
53	40 38	36	99	41 14	36	98	41 50	35	96	42 25	34	95	42 59	34	94
54	39 53	36	98	40 29	36	97	41 05	35	96	41 40	34	95	42 14	33	94
55	39 08	+36	97	39 44	+36	96	40 20	+34	95	40 54	+35	94	41 29	+33	93
56	38 23	36	96	38 59	35	95	39 34	35	94	40 09	34	93	40 43	34	92
57	37 38	36	96	38 14	35	95	38 49	35	94	39 24	34	93	39 58	34	92
58	36 53	36	95	37 29	35	94	38 04	35	93	38 39	34	92	39 13	33	91
59	36 08	36	94	36 44	35	93	37 19	34	92	37 53	35	91	38 28	33	90
60	35 23	+35	94	35 58	+36	93	36 34	+34	92	37 08	+34	91	37 42	+34	90
61	34 38	35	93	35 13	35	92	35 48	35	91	36 23	34	90	36 57	34	89
62	33 52	36	92	34 28	35	91	35 03	35	90	35 38	34	89	36 12	33	88
63	33 07	36	92	33 43	35	91	34 18	34	90	34 52	34	89	35 26	34	88
64	32 22	35	91	32 57	35	90	33 32	35	89	34 07	34	88	34 41	34	87
65	31 37	+35	90	32 12	+35	89	32 47	+35	88	33 22	+34	87	33 56	+34	86
66	30 51	36	90	31 27	35	89	32 02	35	88	32 37	34	87	33 11	34	86
67	30 06	36	89	30 42	35	88	31 17	34	87	31 51	35	86	32 26	34	85
68	29 21	35	88	29 56	35	87	30 31	35	86	31 06	35	85	31 41	34	84
69	28 35	36	88	29 11	35	87	29 46	35	86	30 21	35	85	30 56	34	84

LHA	25° Hc	d	Z	26° Hc	d	Z	27° Hc	d	Z	28° Hc	d	Z	29° Hc	d	Z	LHA
0	74 00	+60	180	75 00	+60	180	76 00	+60	180	77 00	+60	180	78 00	+60	180	360
1	73 59	60	177	74 59	60	177	75 59	59	176	76 58	60	176	77 58	60	176	359
2	73 55	60	173	74 55	59	173	75 54	60	173	76 54	59	172	77 53	60	172	358
3	73 48	60	170	74 48	59	170	75 47	59	169	76 46	59	168	77 45	59	168	357
4	73 39	59	167	74 38	59	166	75 37	59	166	76 36	58	165	77 34	58	164	356
5	73 28	+58	164	74 26	+58	163	75 24	+58	162	76 22	+58	161	77 20	+57	160	355
6	73 14	58	161	74 12	57	160	75 09	57	159	76 06	57	157	77 03	55	156	354
7	72 58	57	158	73 55	57	157	74 52	55	155	75 47	56	154	76 43	54	152	353
8	72 40	56	155	73 36	56	154	74 32	54	152	75 26	55	151	76 21	53	149	352
9	72 20	55	152	73 15	55	151	74 10	53	149	75 03	53	148	75 56	52	146	351
10	71 58	+54	149	72 52	+53	148	73 45	+53	146	74 38	+52	145	75 30	+50	143	350
11	71 34	53	147	72 27	52	145	73 19	52	144	74 11	50	142	75 01	49	140	349
12	71 08	53	144	72 01	51	143	72 52	50	141	73 42	49	139	74 31	48	137	348
13	70 41	51	142	71 32	51	140	72 23	49	139	73 12	48	137	74 00	46	135	347
14	70 13	50	140	71 03	49	138	71 52	48	136	72 40	47	134	73 27	45	132	346
15	69 43	+49	137	70 32	+48	136	71 20	+47	134	72 07	+45	132	72 52	+45	130	345
16	69 11	49	135	70 00	47	134	70 47	46	132	71 33	44	130	72 17	43	128	344
17	68 39	47	133	69 26	46	132	70 12	45	130	70 57	44	128	71 41	42	126	343
18	68 06	46	131	68 52	45	130	69 37	44	128	70 21	42	126	71 03	41	124	342
19	67 31	46	130	68 17	44	128	69 01	43	126	69 44	41	124	70 25	40	122	341
20	66 56	+44	128	67 40	+44	126	68 24	+42	124	69 06	+41	122	69 47	+39	120	340
21	66 19	44	126	67 03	43	124	67 46	41	123	68 27	40	121	69 07	38	118	339
22	65 42	44	124	66 26	41	123	67 07	41	121	67 48	39	119	68 27	37	117	338
23	65 05	42	123	65 47	41	121	66 28	40	119	67 08	38	117	67 46	37	115	337
24	64 26	42	121	65 08	40	120	65 48	39	118	66 27	38	116	67 05	36	114	336
25	63 47	+41	120	64 28	+40	118	65 08	+39	116	65 47	+37	115	66 24	+35	113	335
26	63 08	40	119	63 48	39	117	64 27	38	115	65 05	37	113	65 42	34	111	334
27	62 28	40	117	63 08	38	116	63 46	37	114	64 23	36	112	64 59	35	110	333
28	61 47	39	116	62 26	38	114	63 04	37	113	63 41	36	111	64 17	33	109	332
29	61 06	39	115	61 45	37	113	62 22	37	111	62 59	35	110	63 34	33	108	331
30	60 25	+38	113	61 03	+37	112	61 40	+36	110	62 16	+34	109	62 50	+33	107	330
31	59 43	38	112	60 21	36	111	60 57	36	109	61 33	34	107	62 07	32	106	329
32	59 01	37	111	59 38	36	110	60 14	35	108	60 49	34	106	61 23	32	105	328
33	58 19	37	110	58 56	35	109	59 31	35	107	60 06	33	105	60 39	32	104	327
34	57 36	36	109	58 12	36	108	58 48	34	106	59 22	33	104	59 55	32	103	326
35	56 53	+36	108	57 29	+35	106	58 04	+34	105	58 38	+33	103	59 11	+31	102	325
36	56 10	36	107	56 46	34	106	57 20	34	104	57 54	32	102	58 26	32	101	324
37	55 26	36	106	56 02	34	105	56 36	34	103	57 10	32	102	57 42	31	100	323
38	54 43	35	105	55 18	34	104	55 52	33	102	56 25	32	101	56 57	31	99	322
39	53 59	35	104	54 34	34	103	55 08	33	101	55 41	31	100	56 12	31	98	321
40	53 15	+35	103	53 50	+33	102	54 23	+33	100	54 56	+32	99	55 28	+30	98	320
41	52 31	34	102	53 05	34	101	53 39	32	100	54 11	32	98	54 43	30	97	319
42	51 46	35	102	52 21	33	100	52 54	32	99	53 26	32	97	53 58	30	96	318
43	51 02	34	101	51 36	33	99	52 09	32	98	52 41	32	97	53 13	30	95	317
44	50 17	34	100	50 51	33	99	51 24	32	97	51 56	31	96	52 27	30	94	316
45	49 33	+33	99	50 06	+33	98	50 39	+32	96	51 11	+31	95	51 42	+30	94	315
46	48 48	34	98	49 22	32	97	49 54	32	96	50 26	31	94	50 57	30	93	314
47	48 03	34	97	48 37	32	96	49 09	32	95	49 41	31	94	50 12	30	92	313
48	47 18	34	97	47 52	32	95	48 24	32	94	48 56	31	93	49 27	29	92	312
49	46 33	33	96	47 06	33	95	47 39	32	94	48 11	30	92	48 41	30	91	311
50	45 48	+33	95	46 21	+33	94	46 54	+31	93	47 25	+31	92	47 56	+30	90	310
51	45 03	33	95	45 36	32	93	46 08	32	92	46 40	31	91	47 11	30	90	309
52	44 18	33	94	44 51	32	93	45 23	32	91	45 55	30	90	46 25	30	89	308
53	43 33	33	93	44 06	32	92	44 38	31	91	45 09	31	90	45 40	30	88	307
54	42 47	33	92	43 20	33	91	43 53	31	90	44 24	31	89	44 55	30	88	306
55	42 02	+33	92	42 35	+32	91	43 07	+32	89	43 39	+31	88	44 10	+30	87	305
56	41 17	33	91	41 50	32	90	42 22	32	89	42 54	30	88	43 24	31	86	304
57	40 32	33	90	41 05	32	89	41 37	31	88	42 08	31	87	42 39	30	86	303
58	39 46	33	90	40 19	33	89	40 52	31	88	41 23	31	86	41 54	30	85	302
59	39 01	33	89	39 34	32	88	40 06	32	87	40 38	31	86	41 09	30	85	301
60	38 16	+33	88	38 49	+32	87	39 21	+32	86	39 53	+31	85	40 24	+30	84	300
61	37 31	33	88	38 04	32	87	38 36	32	86	39 08	31	85	39 39	31	84	299
62	36 45	33	87	37 18	33	86	37 51	32	85	38 23	31	84	38 54	31	83	298
63	36 00	33	87	36 33	33	86	37 06	32	84	37 38	31	83	38 09	31	82	297
64	35 15	33	86	35 48	33	85	36 21	32	84	36 53	31	83	37 24	31	82	296
65	34 30	+33	85	35 03	+33	84	35 36	+32	83	36 08	+31	82	36 39	+32	81	295
66	33 45	33	85	34 18	33	84	34 51	32	83	35 23	32	82	35 55	31	81	294
67	33 00	33	84	33 33	33	83	34 06	32	82	34 38	32	81	35 10	31	80	293
68	32 15	33	84	32 48	33	83	33 21	33	82	33 54	32	81	34 26	31	80	292
69	31 30	33	83	32 03	33	82	32 36	33	81	33 09	32	80	33 41	32	79	291

S. Lat. {LHA greater than 180°....... Zn=180−Z; LHA less than 180°........... Zn=180+Z}

DECLINATION (15°-29°) SAME NAME AS LATITUDE

LAT 41°

INDEX

Abbreviations and Symbols, Navigation 70-71
Air Almanac 39, 43
  excerpts 72-82
Almanacs 39
  See also:
    Air Almanac
    Nautical Almanac
Altitude 20, 21, 26, 30, 32-35, 70
  See also:
    Sextant Alt.
    Observed Alt.
    Computed Alt.
    Tabulated Alt.
    Altitude difference
Altitude difference (intercept) 32, 47, 70
Arc 14, 19, 23, 30
Arc to Time Conversion Table
  excerpt 79
Aries, First Point of 56-57, 70
Assumed latitude 44, 45, 70
Assumed longitude 44-45, 70
Assumed position 31-32, 35, 44-45, 70
Astronomical Navigation Tables (H.O. 218) 40, 68
Azimuth angle 21-22, 33, 47, 70
Civil twilight 58-59
  table, excerpt 78
Computed altitude 31-34, 46-47, 70
Conversion Table, Arc to Time
  excerpt 79

Correction to Tabulated Altitude for Minutes of Declination Table 46
excerpt 87
Daylight Saving Time 25, 50, 54, 70
Declination 13, 29, 31, 34, 45, 46, 50, 55, 58, 61, 62, 70
Declination
table excerpts 86,88,89
Dip (height of eye) 35, 70
correction for, table excerpt 82
Dip, correction for 35
table excerpt 82
Evening Civil Twilight 58-59
Table, excerpt 78
First Point of Aries 56-57, 70
Geographical Position 11, 12, 13, 19, 26, 29-31
Great circles 22-23
Greenwich Hour Angle 13-19, 43-45, 52, 56-57, 70
Interpolation Table, excerpt 76
Greenwich Mean Time 14, 24, 42, 43, 51, 54, 55-56, 57, 59, 62, 70
Height of eye (dip) 35, 70
correction for, excerpt 82
H.O. 214—*Tables of Computed Altitude and Azimuth* 39, 40, 61, 68
H.O. 218—*Astronomical Navigation Tables* 36, 40, 68
H.O. 249—*Sight Reduction Tables for Air Navigation* 39, 40, 46-47, 58-59
excerpts 83-89
H.O. 2102-D—*Star Finder and Identifier* 66
Horizon 19-20
Hour Angle 13-19, 56-57
See also:
Greenwich Hour angle

Local Hour Angle
Sidereal Hour Angle
Index error 43, 65, 70
Intercept (Altitude difference) 32, 47, 70
Interpolation of GHA 43
Table, excerpt 76
Key to the diagrams 9
Latitude 13, 30, 31, 55, 70
Local Hour Angle 13-19, 22, 34, 44-45, 46, 57, 59-61, 62, 68, 71
Longitude 12, 15-19, 24, 44-45, 54, 59, 71
Moon Sight 50-52, 62
Morning Civil Twilight 58-59
Table, excerpt 78
Nautical Almanac 39, 68
Navigation Abbreviations and Symbols 70-71
Noon Sight 27, 54-56, 62
Observed Altitude 35-37, 43, 60, 71
Parallax 37
Parallax in Altitude 51, 71
Planet Location Diagram 54
Planet Sights 52, 54, 62
Polaris (Pole Star) Table 61, 62
excerpt 80
Pole Star Sights 29-30, 61, 62
Position circle 26
Position line 25-26
Precession and Nutation Correction 60-61
Table, excerpt 85
Refraction 35, 68, 71
Table, excerpt 81
Rude Star Finder 66
Semi-diameter 36, 52, 60, 71

Sextant 20, 64-66
Sextant altitude 20, 35-37, 43, 51, 64, 71
Sextant corrections 35-37, 64-65
Sextant error
(see also Sextant corrections)
Side error 65
Sidereal Hour Angle 19, 56-57, 71
Sight Reduction Tables for Air Navigation (H.O. 249) 40, 46, 58-61, 62
excerpts 83-89
Sky Diagrams 61, 66, 67
South latitudes 68
Spherical triangle 30-35, 67
Standard Time 24-25
Star Finder and Identifier (H.O. 2101-D) 66
Star identification 61, 66-67
Star Sights 56-61, 62
Sunrise 58
Table, excerpt 77
Sunset 58-59
Table, excerpt 77
Sun Sights 40-50, 62
Symbols and Abbreviations, Navigation 70-71
Tables, navigation 39-40
See also:
H.O. 214
H.O. 218
H.O. 249
Tables of Computed Altitude and Azimuth (H.O. 214) 39, 40, 61, 68
Tabulated altitude 46
Time 24-25, 41-42
Greenwich Mean

Zone
Zone description
Standard
Daylight saving
Watch
True azimuth 21-22, 32, 59, 60, 70
Twilight, civil 58-59
Table, excerpt 78
Watch error 71
Watch time 42, 71
Zenith 19
Zenith distance 20, 29, 30, 62
Zone description 24, 71
Zone time 24, 71